HushHush

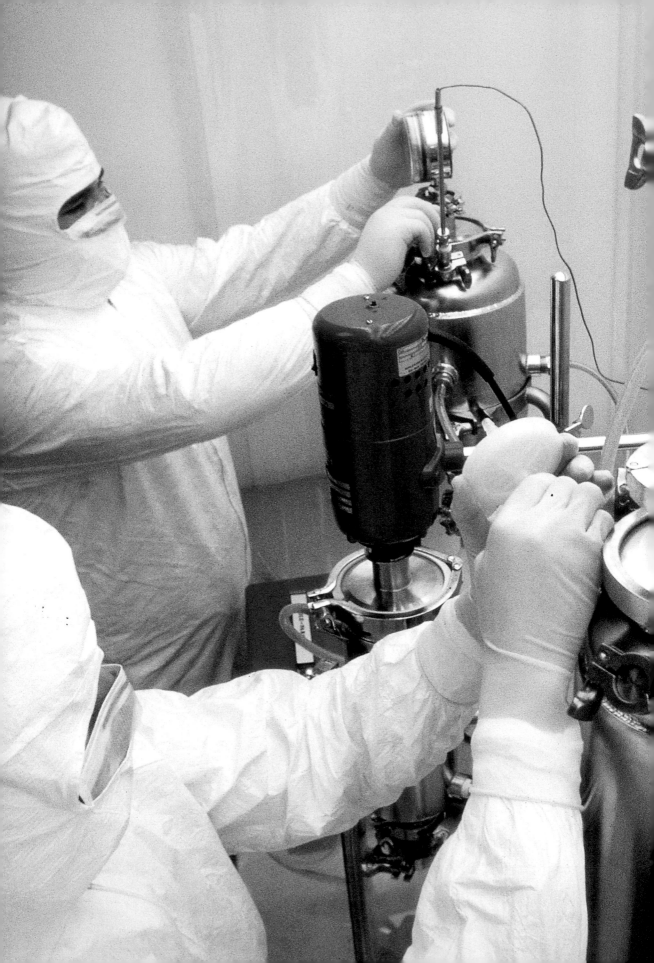

For Gerald Reed, with great affection

HushHush

The dark secrets of scientific research

MICHAEL JORDAN

Foreword by Dominique Lapierre

FIREFLY BOOKS

A FIREFLY BOOK

Published by Firefly Books Ltd, 2003

First printing.

National Library of Canada Cataloguing in Publication Data.

Jordan, Michael, 1941-
    Hush, hush : the dark secrets of scientific research / Michael Jordan.

Includes bibliographical references and indexes.
ISBN 1-55297-607-6 (bound).
ISBN 1-55297-608-4 (pbk.)
    1. Industrial accidents.
    2. Errors, Scientific.
    3. Environmental degradation.
    4. Offenses against public safety.
    5. Corporations—Corrupt practices.
    6. Official secrets.
I. Title.
Q162.J67 2003   363.1   C2002-905271-8

Publisher Cataloging-in-Publication Data (U.S)
(Library of Congress Standards)

Jordan, Michael.
    Hush hush : the dark secrets of scientific research / Michael Jordan ; foreword by Dominique Lapierre. —1st ed.
    [192] p. : ill. , col. photos. ;   cm.
    Includes bibliographical references and index.
    Summary: How scientific developments can sometimes go wrong and how corporations and scientists attempt to cover up the unintended consequences. Includes 40 cases from around the world
    ISBN 1-55297-607-6
    ISBN 1-55297-608-4 (pbk.)
    1. Errors, Scientific—History.   2 Science—History.
I. Title.
509 21 Q172.5E77.J67   2003

Published in Canada in 2003 by:
Firefly Books Ltd
3680 Victoria Park Avenue
Toronto, Ontario
M2H 3K1

Published in the United States in 2003 by:
Firefly Books (U.S.) Inc.
P.O. Box 1338, Ellicott Station
Buffalo, New York
U.S.A. 14205

This book was designed and produced by:
Quintet Publishing Limited
6 Blundell Street
London N7 9BH

Project Editor: Clare Tomlinson
Editor: Jane Merryman
Art Director: Sharanjit Dhol
Designer: James Lawrence
Illustrator: Richard Burgess
Picture Editor: Veneta Bullen
Picture Research Assistant: Renzo Frontoni

Creative Director: Richard Dewing
Publisher: Oliver Salzmann

Manufactured in Singapore by Universal Graphic (Pte) Ltd.

Printed in China by Midas Printing Ltd.

Contents

Foreword

By Dominique Lapierre
Co-author of *Five Past Midnight in Bhopal*

I never cease to thank God for having been born in a world where science, medicine and technology have eradicated most of the scourges and dangers that made our forefathers' lives so fragile. But the extraordinary development in these areas in recent years has also given birth to a whole array of hidden threats, which have become terrible dangers to our survival. The exceptional merit of *Hush Hush* is to reveal some of these dangers. Each and every chapter reads like a frightening detective story. What is at stake in many of these cases is our basic capacity to survive the vicious and often insidious menaces vested upon us by huge and careless business interests. The first chapter of this remarkable work concerns a tragic event to which I have devoted three years of my life reconstructing in a book, *Five Past Midnight in Bhopal*, as co-author with Javier Moro. It explains how an industrial project can begin as a fairy tale and end as a tragedy, killing between sixteen and thirty thousand innocents in one single night and poisoning over five hundred thousand more.

U.S. multinational Union Carbide's project to build a pesticide plant in the heart of India, to help poor Indian farmers eradicate the insects that devastate their crops, was certainly a good idea. But when it applied for a land permit to build the factory so close to the heart of the city, Carbide never told the municipal authority of Bhopal that for producing this particular pesticide they would have to use methyl isocyanate, a gas which is probably the most dangerous substance ever invented by the chemical industry. Carbide could have chosen a location 50 or 100 miles from the heavily populated city of Bhopal. But it needed a lot of water and electricity, and a permanent stock of labor. The immediate outskirts of the city of Bhopal was an ideal place. Anyway, why would anyone object to this choice? Had not a Carbide engineer himself told one of his colleagues: "The plant we are going to build in Bhopal will be as innocent as a chocolate factory!" A chocolate factory which used as its main production component 1000 of liters of methyl isocyanate. Of course, no one at Carbide ignored the abyssal dangers of this gas. It was so volatile that its combination with only a few drops of water or a few ounces of metal dust could prompt an uncontrollable violent reaction. No safety system, no matter how sophisticated, would then be able to stop it from emitting a fatal cloud into the atmosphere. To prevent its explosion, methyl isocyanate has to be permanently kept at a temperature near zero degrees centigrade. Provision had therefore to be made at the Bhopal plant for the refrigeration of any drums or tanks that were to hold it. The Carbide engineers were so sure of their capacity to control the violent temper of methyl isocyanate that they did not hesitate to build their plant a stone's throw away from a belt of slums populated by over 50,000 under-privileged job-hunters.

An alarm system was installed on the different units of the plant to warn workers in case of a sudden leak in the gas circuits. But none of the loudspeakers of this system was turned toward the nearby slums. When I tried to find out why, I was simply told that the management did not want to create panic among the nearby population. Leaks occurred frequently. More disturbing than this is the fact that the windsock mounted on the top of the plant's main unit could not be seen from the adjacent *bustees* (slums). This windsock provided an essential piece of information to the employees working on the site; it told them in which direction they had to run away in case a major gas leak occurred. This windsock was not visible from the slums. And again, in its efforts to stay hush-hush on the dangers of the plant, the management did not bother to provide the slums with their own windsock. It could have saved hundreds, perhaps thousands, of lives on the night of December 2, 1984. Instead of running away from the lethal cloud, which the wind pushed behind them to eventually catch up with them, Bhopal's inhabitants would have run against the wind. All Carbide workers present on the plant that night were able to save their own lives thanks to this piece of cloth flying in the air.

Dominique Lapierre outside the abandoned Union Carbide pesticide plant in Bhopal.

The Carbide coverup about the dangers of methyl isocyanate gas took on criminal proportions on the night of the tragedy. Forty-five minutes after the explosion, the plant manager, J. Mukund, responded to frantic calls he was receiving that: "The gas leak just can't be from our plant. The plant is shut down. Our technology just can't go wrong. We just can't have such leaks." A few minutes later, Dr. L.D. Loya, the plant's chief medical officer, told the doctors who were calling for help from all corners of the city: "There is nothing to do except ask patients to put a wet towel over their eyes. The gas is nonpoisonous." Eighteen years after the tragedy, Carbide has still not revealed the exact composition of the gas that killed and poisoned so many. This scandalous coverup has prevented doctors from providing specific treatment for the victims.

Once it realized the enormity of the disaster it had created, Carbide adopted a mode of "crisis management" focused on "damage limitation" to the corporation. The most blatant effort to promote this policy was to shift the blame for the tragedy from the company to an individual. Four months after the accident, the vice president of the agricultural division of Carbide's Indian subsidiary called a press conference in Bombay to announce that the tragedy had not been due to an accident but to sabotage. He based his statement on an enquiry carried out by the team of engineers sent to Bhopal the day after the disaster. According to this enquiry, a worker had deliberately introduced a large quantity of water into a pipe connected to one tank full of methyl isocyanate gas. This worker, who remained nameless, had supposedly acted out of vengeance after a disagreement with his superiors. To support this theory, the investigators had relied on the discovery of a hosepipe close to the tank and upon the doctoring of log book entries made by the shift on duty that night. The report made no mention of the fact that none of the factory's security systems was functioning at the time of the accident.

I and the co-author of *Five Past Midnight in Bhopal* found the man Union Carbide had accused. We talked to him at length. The man in question is a young operator named Mohan L. Verma. It is our deep-seated conviction that this father of three, who was well aware of the dangers of methyl isocyanate, could not have perpetrated an act to which he, and a large number of his colleagues, were likely to fall victim. Mohan L. Verma's innocence was, moreover, never truly disputed. No legal proceedings were ever instituted against him. Today, he lives quite openly two hours by car from Bhopal. If the survivors of the tragedy had had the slightest suspicion about him, would they not have sought vengeance? As it was, no one in Bhopal, or elsewhere, took the charge seriously.

Eighteen years later, Union Carbide, now owned by Dow Chemical Corporation, is still evading its responsibility for history's biggest industrial disaster; it is still hiding behind this sabotage tale. In spite of an Interpol warrant served against him at the request of the Indian authorities in 1992, Warren Anderson, the chairman of Carbide at the time of the accident, has never been brought to court. Is there a chance that this man is indirectly responsible for the death of six times more victims than perished at the New York World Trade Center? Will he ever face justice? Or will we have to continue to believe that corporations can get away with murder? Thousands upon thousands of the victims of this man-made tragedy still suffer in their flesh today; they would like to know if justice will be served before they die.

The Bhopal tragedy is just one of the 40 scientific coverups detailed in *Hush Hush*. Read on and learn about the secrets held by governments and businesses alike.

Royalties from *Five Past Midnight in Bhopal* support a gynecological clinic in Bhopal that treats destitute victims of this tragedy. **www.cityofjoyaid.org**

Introduction

We are, by nature, secretive creatures. Hiding certain things away from the eyes and ears of others is one of the quirks of human behavior. In this we are by no means alone in the animal kingdom—birds often hide their nests in the densest parts of trees, and dogs covertly bury their bones.

According to the earliest recorded evidence discovered, the desire for secrecy seems to have been ingrained in us. Archeological discoveries from the Ice Age appear to confirm that our Neolithic ancestors created special closed sanctuaries in which their most important rituals were held. Caves in southwest France, decorated more than 15,000 years ago with elaborate paintings and esoteric designs, are often virtually inaccessible, even to this day, and we can guess that the intention was to keep certain activities known only to those holding the reins of authority. The desire for power is indeed one of the fundamental values of secrecy because, through its application, a limited number of people can generate fear of the unknown in the minds of the majority, thus better exercise control.

Much of the secrecy we encounter and follow in our lives today is of no great consequence to the world at large and we often maintain it to protect our personal interests. Just like the dog with its bone or the bird in the bush, we lock our front doors to prevent unwanted intrusion and keep our bank account numbers and the passwords on our computers to ourselves for obvious reasons.

Secrecy, however, can always be misused and abused. It can be made a device to screen activities from public gaze because of a wish to hide away conduct that would be disapproved of—that could perhaps become the object of sanctions were it to be more widely advertised or held open to inspection. Nowhere can this be truer than in matters of scientific secrecy. Until the 1960s, most scientific discovery and progress remained bathed in a somewhat rosy light. We trusted scientists; they were honest people working on our behalf to make our lives and our world safer and richer; they eliminated many of our life-threatening diseases and made our cars, ships and airplanes safer and more efficient. But in today's world, we are a little more cynical, and with good reason. This is not to say that scientists are corrupt as a breed, but history has shown us that not all scientists are altruistic angels either. The

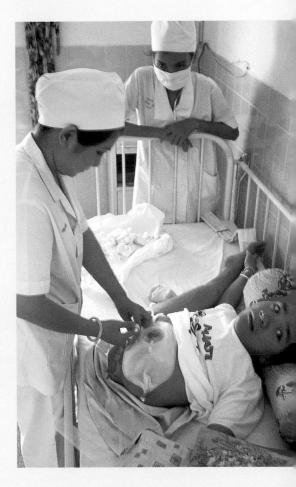

Unknown to most people outside the manufacturing companies, the Vietnam War defoliant Agent Orange was contaminated with lethal dioxins. Military personnel and civilians suffered alike.

↑

Conflicting scientific data about global deforestation and its potential effects has been exploited by both sides in the controversy.

scientific establishment can be bought and its integrity can be corrupted, although money is not always the driving motive. Scientists can be persuaded to use their expertise in a particular, not necessarily aboveboard, way out of a sense of political or national loyalty or for ideological reasons.

Secrecy in science may not always be strictly ethical, but neither does it necessarily result in harm to others. In some instances, however, the desire to cover up scientific activities is seen—when exposed to scrutiny —to be blatantly against the interests of society, and this can result in civil actions and even criminal charges being brought against those conspiring to maintain the secrecy.

When it is misused or abused, such secrecy often reflects an unreasonable desire for profit or a wish to conceal errors of judgment. Governments and major corporations alike display a willingness to hold on to as much knowledge as they can, on an exclusive basis, because it gives them the advantage of making use of scientific discoveries. Perhaps the most classic illustration of this during the last century has been the Manhattan Project in the U.S. During World War II, the United States learned that German scientists were working on a program developing the use of atomic fission with the intention of producing a nuclear bomb. So alarming was this prospect, and so essential was the need for America to gain the advantage in the arms race, that it orchestrated a vast weapons program innocuously called the Manhattan Project, which involved thousands of scientists and technicians in a level of previously unparalleled secrecy.

In the past, when issues of civil rights were not as prominent in many countries as they are today, it was comparatively easy to maintain secrecy. Even now, in the twenty-first century, secrecy is still widespread under certain regimes. Perversely, however, the desire for openness and fairness has, in some respects, actually encouraged secrecy. We can see instances in which governments and corporations have done their utmost to keep certain information under cover because of the perceived threat of litigation. These days, where damages are sought for malpractice or negligence that result in suffering being inflicted on large numbers of third parties, the compensation sums ordered by the courts can run into billions of U.S. dollars. Such punitive judgments have already bankrupted a number of large organizations.

Corporations have increasingly turned to legal devices in order to protect themselves. Among these is a reliance on the confidentiality of the attorney-client privilege, although the loophole in some nations is slowly being closed. A fraud-crime exception to such a right is now on the statute books of several countries, which allows documents to become public. In a case against tobacco companies in the United States a few years ago, a judge in Minnesota declared that the defendants had "blatantly abused" attorney-client privilege in withholding information that was potentially incriminating and ordered that more than 30,000 documents be turned over to the court. The state attorney for the

Whaling countries, including Norway and Japan, are criticized for their attitudes to the conservation of much depleted whale stocks.

prosecution commented at the time that a "40-year wall of fraud and secrecy has been breached...this has far reaching implications for the defendants and their officers and lawyers." Nonetheless, large corporations can tie up proceedings for years in order to delay rulings and postpone facing liabilities.

Time is sometimes of the essence in revealing secrecy. Coverups, particularly by governments and military authorities, are often introduced out of concern for national security. Many incidents from World War II have only recently come to light because of the declassification of millions of confidential and secret documents half a century later. Fifty years may seem a long time to hide away sensitive wartime information, but it pales in comparison to the timescale of some other illustrations of secrecy. Although clearly not a scientific issue, it is only during the last decade or so that the Vatican archives of the Roman Catholic Church's part in the trials of the Papal Inquisition more than 500 years ago have been opened to inspection, so sensitive is the nature of some of the material.

Any book reviewing scientific secrecy in the twenty-first century inevitably has its limitations, because the hush-hush cases that have come to light in the courts and through the media do not, of course, represent the full story. If a coverup is effective, it remains clandestine and we never get to hear about it. Unfortunately, we can only examine those cases on which the whistle has been blown.

REFERENCES

Document: potential smoking habits of 5-year-olds reviewed. *CNN Interactive.* March 7, 1998.
www.cnn.com/US/9803/07/minn/tobacco/

Among the most extraordinary coverups of the 20th century has been the Manhattan Project in the USA. The extent of radiation damage from the nuclear test programs of the 1950s and 1960s is still shrouded in secrecy.

1 The Secret of Success

1   The Secret of Success

In the age of big business, when international conglomerates have annual turnovers bigger than the gross domestic products of some nations, the power and influence of some companies' boards of directors can equal or surpass that of the regulatory bodies that claim to hold them accountable. Companies answer primarily to shareholders, whose main interests lie in profitability. In case after case, this benchmark for commercial success has led to disastrous consequences.

⬆

The residents of Bhopal, in the Indian state of Madhya Pradesh, were the unwitting victims of the world's worst industrial accident. Lax safety standards and poor staff training led to the poisoning of a whole town and the death of thousands of people. The owners of the plant have never fully cleaned up the area and even now the residents must drink polluted water.

It is an unfortunate fact of life and human nature that scientists and technicians, physicians and engineers, like any other breed of professional, can be bought. It takes a brave employee to force the publication of a report or exposé that runs contrary to the interests of the company he or she works for. We have seen, in recent years, well-publicized illustrations of the punitive consequences individuals face when they blow the whistle on their management superiors concerning matters of public interest and safety. You may possess incontrovertible evidence that the company you work for is exposing its factory-floor employees to health risks, but go public with your information and it is likely that you will find yourself out of a job. Allege that the aircraft you maintain, the rail track you inspect or the road tunnel you build possesses a fundamental design fault or a weakness in its construction and your own safety can be put at risk.

In many countries, big business frequently finances government coffers, whether at a local or national level. Industrial giants have politicians from all over the world in their pockets, either through giving individual (and often influential) politicians highly paid directorships or through donations to political parties. The true motivations for these contributions are rarely revealed, though they can range from the diversion of potentially awkward questions and investigations to getting unfair advantages when bidding for government contracts. This incestuous relationship between government and business prevents many secrets being revealed. Nationalized industries are not necessarily any less culpable than their commercial counterparts, either, as impossibly high financial or service-led targets can force them to cut corners at the expense of public interest.

Even when malpractice is exposed or a disaster comes to the attention of the world, it does not necessarily mean that the walls of secrecy are demolished. At this juncture, the legal specialists take over and use such devices as attorney-client privilege. In this manner, they can evade or postpone the disclosure of documents that might otherwise provide damning evidence against the defendants they serve. Lawsuits brought by plaintiffs suing for damages against big business can be held up almost indefinitely by the use of such legal machinations.

One of the tricks employed to avoid liability appears to be the timely splitting off of one section of a conglomerate known to be at risk of facing either civil or criminal court action. An affiliated commercial organization is formed, often based in another country, over which the parent company may legally disclaim responsibility. Thus, a multinational is able to protect its shareholders from the worst effects of punitive damage claims, some of which can result in company insolvency and even bankruptcy. The recently formed affiliate is sacrificed as the scapegoat and the offending management in the core business walks away comparatively unscathed.

But trade secrecy is an essential part of how businesses operate, so in many cases, where no personal or environmental harm is inflicted, it is justifiable. If a company produces a product with features unique from its competitors, be it anything from a car to a dishwashing liquid, it would be commercial suicide to disclose all privy details of the design, the ingredients or the manufacturing process. Patent and copyright laws protect these issues. But what is unacceptable is a conscious decision by a company to keep details of damaging or unethical practices concealed from the general public. It may be easier and more cost-effective to produce a car with a vulnerable fuel tank, an aircraft with a suspect system or a herbicide that is not adequately tested; and it may be seen as profitable to run an industrial plant in which safety features are pared down below an acceptable level. But such decisions are morally inexcusable. The people who make such judgments must realize the immorality of what they are doing, otherwise why would they feel the need for secrecy?

Research carried out at the University of Zurich has linked electronic radiation damage to brain cells with the use of cell phones. Manufacturing companies continue to deny that such effects exist.

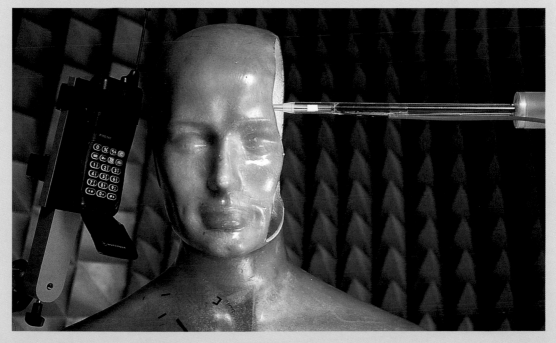

1.1 The Bhopal Chemical Disaster

When a toxic chemical cloud escaped from the Union Carbide plant at Bhopal in northern India, the effects killed many thousands of people on the first night. Thousands later succumbed, yet the company is alleged to have withheld vital medical information on the grounds of "trade secrecy."

During the night of December 2, 1984, the hospitals in Bhopal, a city in the Indian state of Madhya Pradesh, began to receive the victims of what was to become the world's worst industrial accident. The medical staff was besieged by large numbers of dying people suffering from massive reactions to a cloud of gas that had erupted from a nearby chemical plant owned by the Union Carbide Corporation. Yet when doctors called the company's medical officer asking how to treat their patients, the response was simply, "All you need do is to wash with water." From that terrifying night the death toll ran into the thousands.

History of Union Carbide India

The Union Carbide factory in Bhopal was set up in 1969 in order to produce the pesticide Sevin for the new Indian market. For the first few years, the residents of the city welcomed the multinational corporation as it provided much-needed employment for so many people. But doubts began to surface and rumor began to spread following a few minor accidents and a more serious one that resulted in an employee's death.

At first, the Bhopal plant produced Sevin from a multitude of imported chemicals. One of the major chemicals used in the manufacture of the pesticide is methyl isocyanate (MIC), a highly dangerous substance that can break into an uncontrollable violent reaction with just a few drops of water or a little metal dust. For the first few years of production, up until 1979, Union Carbide India imported MIC from the USA, but after this time it began to manufacture its own. By this time, the Indian subsidiary was struggling financially, as the Indian market for pesticides had failed to take off as predicted. Despite the design of the MIC unit being based on Union Carbide's modern plant in West Virginia, corners were cut in the materials used to construct it, the safety mechanisms and the metering gauges.

With sales in decline, the corporation wanted to save money. Between 1980 and 1984, the number of staff in the MIC unit was halved. There was a high turnover and new employees were not trained to the same level as previous intakes. The plant stopped producing Sevin constantly and began manufacturing intermittently when there were the sales to support it. Huge vats of chemicals lay dormant and large stocks of MIC remained in their tanks. Safety was lax as the inadequately trained operators believed a dormant plant was not a danger. Safety systems were shut down to save money; the MIC refrigeration unit was closed to save electricity bills, despite the fact that MIC has to be kept at zero degrees Celsius at all times.

The fateful night

The disaster occurred during routine maintenance of the plant that manufactured MIC. Leaking valves and corroded pipes triggered a runaway exothermic reaction that resulted in a massive 40 tons of this

The underground three-flanged stainless steel tank (lower left foreground) at Union Carbide's Bhopal factory from which poisonous gas leaked. The tank was uncovered by officials probing the tragedy.

More than 120,000 survivors of the Bhopal disaster still require medical attention and many continue to die from exposure-related illnesses.

compound, as well as hydrogen cyanide and a cocktail of other chemicals, blasting out in dense clouds that rolled over half a million people, the closest of which were in the city's three bustees, makeshift housing estates inhabited by the city's poorest people. Within minutes, thousands upon thousands of people were suffering the horrifying effects of the poisonous gases, many of whom would die that very night. The toxins attacked the victims' eyes, lungs, kidneys, livers, intestines, muscles, brains, reproductive and immune systems. People gasped for breath and staggered around on the verge of going blind. A spokesman for a leading pressure group, the Bhopal Group for Information and Action (BGIA), put the number of people who died immediately at not less than 8,000, although later reports suggest it is far more as the city's hospitals could not count the dead fast enough. It was later calculated by an Indian government agency, the Council of Medical Research, that more than half a million individuals received poisons in their bloodstream resulting in damage to almost all of the systems in their bodies.

Need to know basis

The BGIA was established after some time went by and the issue of compensation had not been satisfactorily resolved. Originally, the Indian government had claimed more than $3 billion on behalf of the victims, but after years of litigation, Union Carbide offered just $470 million in a civil settlement, less than one-seventh of the amount. Faced with threats that the company would stall the case endlessly, the Indian government accepted the offer in February 1989. In 1991, the Indian Supreme Court upheld the decision while allowing the criminal case to be reopened. Union Carbide stated that the amount paid represented a final settlement of all Bhopal civil litigation. Yet people continue to die every month from exposure-related illness, and over 120,000 survivors still require medical attention. Union Carbide is said to have withheld vital information on the exact nature of the gas cloud that spewed from the plant, and its potential effects on the body, on the grounds that such information would breach its right to "trade secrecy." Doctors still do not know the proper treatments and in consequence, many of the drugs being used in local clinics may be causing more harm than good.

Taken to court

In November 1999, against this background and in the face of claims that over 95 percent of claimants had received only derisory sums in compensation (less than $3,000 in the case of death), plaintiffs, including BGIA, filed a lawsuit in New York against the Union Carbide Corporation and its former chief executive officer, Warren Anderson. The suit cited violation of international law and the fundamental human rights of victims and survivors. It charged that "the defendants are liable for fraud and civil contempt for their total failure to comply with the lawful orders of the courts of both the United States and India."

Thousands of inhabitants of Bhopal fled their homes after the Indian government allowed technicians to reopen facilities at the Union Carbide pesticide plant just days after the accident. Union Carbide's management said that the best way to dispose of the lethal MIC that remained in the tanks was to temporarily restart production.

In August 1999, the Union Carbide Corporation was bought by the Dow Chemical Company, making the conglomerate one of the biggest in the chemical industry. Among Dow's statements on its responsibilities to the public and the environment has been the recent assertion that "Dow is in the business of applying science and technology to improve the quality of life around the world... whenever Dow has experienced a serious or fatal accident, we have communicated openly and honestly with our people, our communities and local authorities explaining why the accidents occurred and how we can prevent them happening again." Sadly, the people of Bhopal may not wholly see eye-to-eye with this sentiment.

Getting away with murder

Those responsible for this catastrophe have never been brought to justice. Union Carbide did not send a disaster relief team to Bhopal; it has never disclosed the full compositional details of the deadly gas to the medical teams who have treated the survivors; it has not fully compensated the survivors and the families of the dead; and it has not cleaned up the polluted factory to allow current residents of the city to rebuild their lives and their communities free of toxic hazards.

At the time of writing, the factory at Bhopal lay abandoned, a decaying epitaph to gross mismanagement or, as the lead counsel for the plaintiffs, Kenneth McCallion, put it, a "demonstration of a reckless and depraved indifference to human life in the design, operation and maintenance of the Union Carbide of India Ltd (UCIL) facility." The civil suit is still pending in the United States. Its claims also concern the contamination of soil and groundwater from the toxic chemicals, a situation which Union Carbide is said to have responded to inadequately by merely covering the area with topsoil. A spokesman for the action groups has revealed that Dow "keeps bringing out this public relations line about how they're not responsible and Union Carbide is a separate entity. While it is a separate entity in form, it still is a wholly owned subsidiary of Dow Chemical." The Bhopal case is by no means the only illustration of multinational companies attempting to avoid direct public liability by arguing that mistakes are the responsibility of their subsidiaries.

October 12, 1984: Union Carbide Chairman Warren Anderson tells the media that the victims of history's worst chemical disaster will be fairly compensated but his firm is not criminally responsible for the catastrophe.

On the 10th anniversary of the Bhopal catastrophe, thousands of the city's inhabitants took to the streets to protest about the lack of forthcoming compensation and the failure of the cleanup operation to make their city safe.

REFERENCES

Brockley, R. Corporate Profile Dow. 1. *The Menace from Midland. Bhopal (Union Carbide Corporation)*. Undated.
www.bhopal.net/corpprofiledow.html

Incident Review. *Bhopal (Union Carbide Corporation)*. Undated.
www.bhopal.com/review.htm

Jones, S. Survivors pressure Dow on Bhopal aftermath. *The Washington Post*. May 7, 2002.
www.washingtonpost.com/wp-dyn/articles/A43761-2002May7.html

Noronha, F. Union Carbide sued in U.S. for 1984 Bhopal gas release. *Environment News Service*. November 16, 1999.
http://ens.lycos.com/ens/nov99/19999L-11-16-02.html

1.2   Asbestos

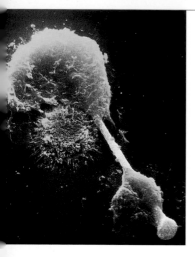

Microscopic 'needles' of asbestos, virtually indestructible, destroy cells in the lungs in a process that triggers the development of mesothelioma, a potentially fatal cancer associated with asbestosis.

No one knows just how many people around the world have died or are suffering from a type of lung cancer medically referred to as mesothelioma, which is almost exclusively caused by exposure to asbestos.

Time is of the essence

One of the problems that has bedeviled victims of the lung disease asbestosis, which in many cases leads to mesothelioma, is the long delay between exposure to asbestos dust and the development of lung cancer, often between 15 and 25 years. Asbestosis can also require only a small exposure to the dust. Scientific evidence linking lung cancer and occupational work with asbestos began to emerge in the late 1930s after American and British asbestos producers sponsored research by the Saranac Laboratory in New York State, under the direction of Dr. Leroy Gardner. Initially, Gardner's findings suggested that asbestosis was curable, but later experiments on laboratory mice made him suspect that a link existed between asbestos dust and incurable mesothelioma, and that further research was required.

The sponsors withheld these findings from publication and one of the major British asbestos mining companies, Turner and Newall, rejected the Saranac findings while promoting a report they had commissioned separately from the British Postgraduate Medical School, concluding that asbestos posed no danger as a cancer-forming agent. Nazi Germany first openly accepted the link between asbestos inhalation and mesothelioma in 1943. Asbestos producers elsewhere continued with their efforts to suppress information that would damage their interests, and in 1951 they succeeded in pressuring Gardner (and the Saranac Laboratory) to withdraw a critical chapter on lung cancer from the draft of a book he was publishing about asbestosis.

The history

In 1953, John Knox, a doctor working with Turner and Newall, joined forces with a highly respected expert in cancer statistics, the late Professor

Asbestosis affected lungs (right) show large blurs on X-rays in comparison to relatively healthy lungs (left).

The mining of asbestos continues in certain parts of the world. Here, a worker at an asbestos mine in Brazil, which is exploited by France's Saint Goban group, holds out asbesto fibers.

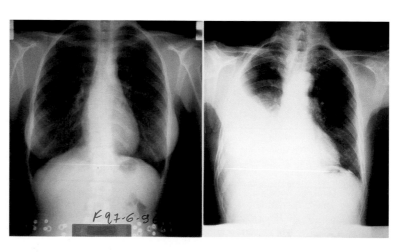

Mesothelioma's most famous victim was Hollywood actor Steve McQueen. During a stint in the U.S. Marines, as well as during countless temporary jobs on construction sites and shipyards, McQueen inhaled asbestos. Following diagnosis of the disease, which in his case was untreatable by orthodox methods, the actor went to Mexico for radical experimental treatment, only to suffer a fatal heart attack not long after, in 1980.

Sir Richard Doll, to produce a new damning report showing that workers in the asbestos industry were at 10 times the risk of contracting mesothelioma compared with the rest of the population. The company refused to release the report, claiming it was inaccurate, and threatened litigation against Doll if he published independently. Doll ignored the threats and submitted the report to the British Journal of Industrial Medicine in 1954. Although a representative of Turner and Newall placed overt pressure on the editor to reject it, the article was published.

The avoidance of responsibility for the dangers of inhaling asbestos dust is finally beginning to catch up with the culprit manufacturers. In October 2000, in the face of massive liabilities, the U.S. insulation manufacturer Owens Corning filed for bankruptcy because of its inability to meet a deluge of lawsuits from asbestos victims. A year later, the U.S. automotive parts maker Federal-Mogul, which had bought Turner and Newall three years earlier for $1.5 billion, went down the same route. In December 2001, U.S. oil services company Halliburton was ordered to pay $30 million in damages relating to asbestos exposure. In 2002, media giant Viacom was being sued by close to 130,000 people with asbestos-related claims against its industrial subsidiaries.

Taking sides

One must question why the British government took almost half a century to recognize what the Germans had first discovered during World War II. Why did it take so long for subsequent British governments to recognize that mesothelioma, when contracted in conjunction with asbestosis, is a disease for which compensation is legally justifiable? It also seems reasonable to ask why, in December 2001, a U.K. court threw out compensation claims relating to exposure in the 1960s, resulting in demands for a House of Lords hearing. Whose interest was being protected? Surely it was not that of the victims who had worked in the asbestos industry. The position is summed up in a terse comment from the GMB Union's Director of Health, Nigel Bryson: "This appears to be a political decision to save asbestos companies and insurers money while victims die in poverty and pain."

On May 16, 2002, in a landmark decision, the five English law lords opened the door to proper compensation for sufferers of asbestos-related cancer when they decreed that three plaintiffs in a test case had the right to claim proper compensation. Made up of one surviving mesothelioma sufferer and two widows who lost their husbands to the disease, the decision has been heralded as the most significant in the history of industrial disease compensation and is likely to cost the U.K. insurance industry trillions of pounds sterling.

REFERENCES

Clark, A. Asbestos claims drain Federal-Mogul. *The Guardian*. October 2, 2001.
www.guardian.co.uk/business/story/0%2C3604%2C561655%2C00.html.

Insurers play down asbestos ruling. *BBC News*. May 17, 2002.
http://news.bbc.co.uk/hi/english/business/newsid_1992000/1992499.stm.

Unattributed article. Owens Corning files Ch. 11. *CNN Money*. October 5th 2000.
http://money.cnn.com/2000/10/05/companies/owens_corning/index.htm

Supreme Court accepts asbestos claims case. *Associated Press*. November 4, 1996.
http://centralohio.thesource.net/Files3/9611048.html

In 1995, when extensive use of asbestos was discovered in the former seat of the European Commission in Brussels, the Berlaymont building had to be abandoned while workers in full protective clothing removed the lethal material.

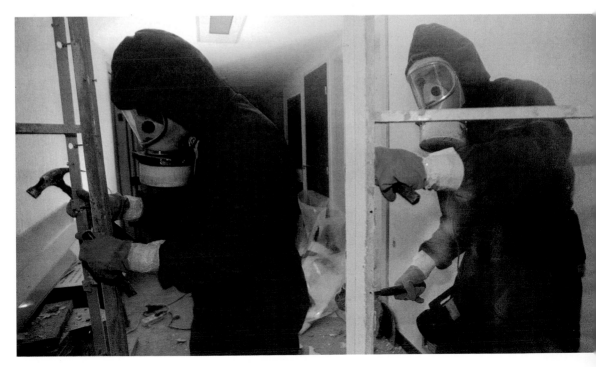

1.3 Tobacco, Health and Lies

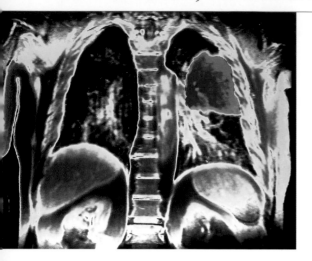

This MRI scan of the chest of a patient suffering from tobacco-related lung cancer reveals a large malignant tumor in the left lung (colored red, upper right).

When a multi-million-dollar industry is confronted with unassailable medical evidence that its products are killers, the prognosis for the viability of that industry is not good. But when evidence comes to the fore suggesting that the tobacco manufacturers were aware of the dangers and continued to advertise and market their brands regardless, the matter takes on darker colors. Anti-smoking campaigners Ross Hammond and Andy Rowell, working on behalf of pressure groups, including the Campaign for Tobacco-Free Kids in the U.S. and Action on Smoking and Health in Britain, published a report in May 2001 claiming to show that denial, deceit and obfuscation are the major tools of the tobacco trade. In almost every area they have touched, the cigarette makers have said one thing to the public and governments, but in the privacy of their boardrooms, laboratories and PR company offices, they have said quite another. The great public controversy around smoking is not the result of honest people who simply have different views, but a carefully and expensively orchestrated campaign by tobacco companies determined to confuse the public.

Is there any justification for such forceful statements? Apparently the answer is yes. A BBC news report of March 1998, for example, reveals evidence of a confidential internal memorandum sent from the research department of the British cigarette manufacturer Gallaher Ltd. to the company's managing director in April 1970. The memo concerned the conclusions drawn from research carried out in the 1960s using beagle dogs as smoking "guinea pigs." The laboratory test animals were forced to inhale cigarette smoke and almost all suffered adverse health effects. The Gallaher research team confirmed "beyond reasonable doubt" that there was a link between cigarette smoke and the incidence of lung cancer that was found to develop in the beagle. They advised of the high likelihood that cigarette smoke inhalation encouraged carcinomas in humans. This information was disclosed on the instruction of a U.S. court during proceedings against tobacco companies.

Speaking out

Over decades, various concerned individuals within the industry had tried and failed to bring information about the danger of tobacco products into the open. In 1988, Jeffrey Wigand, a biochemist who had won a senior research and development position with a subsidiary of British American Tobacco (BAT), is alleged to have found that company lawyers were restricting access to scientific papers and preventing announcements of research results into ways of developing safer cigarettes. The argument appears to have been that the publication of such research might expose tobacco products as being unsafe and leave tobacco companies more open to litigation.

In 2000, the House of Commons Health Committee in the United Kingdom set about reviewing the PR strategies of tobacco companies including Philip Morris, Gallaher, British American Tobacco, Imperial Tobacco and Japan Tobacco. The committee summarized its findings as follows:

> *It seems to us that the companies have sought to undermine the scientific consensus until such time as that position appears ridiculous. So the companies now generally accept that smoking is dangerous (but put forward distracting arguments to suggest that epidemiology is not an exact science, so that the figures for those killed by tobacco may be exaggerated); are equivocal about nicotine's addictiveness; and are still attempting to undermine the argument that passive smoking is dangerous.*

The black market

In the last few years, a more sinister aspect of tobacco company activity has come to light. According to Raymond Bonner and Christopher Drew, writing in the New York Times in August 1997, worldwide smuggling of cigarettes nearly tripled in volume during the 1990s. Independent research indicates that 25 percent of all cigarettes sold overseas pass through smuggling rings.

Manufacturers hold up their hands claiming that they neither condone nor encourage cigarette smuggling, but some of those more closely involved in commercial cigarette dealership tell a

 Hollywood movie *The Insider*, starring Russell Crowe and Al Pacino, tells the story of Dr Jeffrey Wigand. The former tobacco industry executive blew the whistle on his former employers, Brown and Williamson, for hiding the knowledge that nicotine is addictive and that the products they produce contain additives known to increase the danger of disease.

different story, suggesting that it is not uncommon for companies to sell massive quantities of leading brands to traders and dealers acting as conduits to the smugglers. The New York Times article cites an interview with one of the major European traders, Swiss-based Michael Haenggi, who confirmed that he has been buying popular brands from the second largest American manufacturer, R. J. Reynolds Tobacco Company, for many years. In 1996 alone, his company handled sales worth $100 million and he admitted that he had resold much of the stock to smugglers in Spain. This figure, however, is only a drop in the bucket. The journalists indicate that organized crime syndicates in Italy, smuggling cigarettes originating from the biggest American manufacturer, Philip Morris Companies Inc. and passing through Swiss dealers, were taking in some $500 million a year.

The defense of the cigarette companies is that governments are to blame—the fundamental causes of cigarette smuggling are high duties and taxes. They should not be responsible for what happens to their

CIGARS. The Big New Trend in Cancer.

Anti-smoking advertising has become as sophisticated as the tobacco companies' campaigns. As shown in this advert, one large cigar is said to be the equivalent of smoking a whole pack of cigarettes.

products once sold and, in any event, they do not knowingly sell to dealers who supply smugglers. Yet some dealers would not agree with this last assertion. Interviewed by Bonner and Drew, a retired dealer of Philip Morris products in Switzerland Corrado Bianchi said, "If the companies say they do not [know that certain sales are destined for the contraband route], it's a lie...of course they know." At least one senior cigarette company official, the senior vice president of a European Philip Morris subsidiary, is reported to have agreed that contraband activities go on. Switzerland, allegedly, does not regard cigarette sales by dealers to smugglers as a criminal offense and European investigators complain that Switzerland is no more willing to cooperate than are the tobacco companies.

Smuggling has coincided with a fall in at home sales, resulting from widespread publicity about the adverse medical effects of smoking. The contraband route also facilitates the illegal import of cigarettes by the smugglers where there are legitimate sales obstacles such as high taxes and import restrictions. Manufacturers earn the same revenue whether their products are sold legitimately or as contraband, and smuggling is seen by many as a means of maintaining sales when the more traditional market has started to dry up.

It is reported that cigarette smuggling is rife in China, for example, where there is a huge potential market, said to be the "ultimate salvation" for the tobacco companies. The Chinese authorities are concerned not only about health problems that could come from an aggressive marketing assault from the West, but also about the vulnerability of the state-run tobacco company that generates large amounts of government revenue. Foreign manufacturers are discouraged from selling their goods in China. Estimates indicate, however, that between 70 billion and 80 billion cigarettes, half of them originating with British and American tobacco companies, arrive in China each year through illegal trafficking.

Leopards don't change their spots

The tobacco industry claims to have reformed, and maintains that it should not be judged on past behavior. Yet those critical of its activities point to such examples as the Chinese situation, arguing that in reality the industry has not changed at all and cannot be trusted. Litigation currently underway in the United States on behalf of lung-cancer victims of smoking is finally beginning to take the lid off 50 years of secrecy and lack of accountability.

China's tobacco industry generated $8.6 billion in profits and taxes in 1995, but the anti-smoking lobby claims that smoking costs the country more in early deaths, lost productivity and medical costs than it gains in tax revenue.

One of the most damning of recent comments appeared in the *Lancet*, in April 2002, describing how tobacco industry documents that were exposed in recent court cases indicate that increase in smoking-related deaths is regarded as a "by-product" of achieving market-expanding objectives. Among these objectives has been tough political lobbying to dissuade the European Union from passing a directive to ban tobacco advertising and sponsorship.

Tobacco companies in the U.S.A. and the U.K. have been accused of trying to undermine scientific evidence about the health risks of smoking. Here, senior executives face the U.S. House of Representatives' Health and Environment Subcommittee in Washington, D.C.

REFERENCES

Bonner, R. and C. Drew. Cigarette makers are seen as aiding rise in smuggling. *New York Times*. August 26, 1997.

Hammond, R. and A. Rowell. Trust us: we're the tobacco industry. Campaign for Tobacco-Free Kids (USA); *Action on Smoking and Health (UK)*. May 2001.
http://tobaccofreekids.org/campaign/global/docs/searching.pdf.

House of Commons (UK) Health Select Committee, Second Report Session 1999–2000. The tobacco industry and the health risks of smoking. *The Stationery Office*. June 2000.

Incriminating evidence against tobacco company. *BBC News Online UK*. March 15, 1998.
http://news6.thdo.bbc.co.uk/low/english/uk/newsid_65000/65.676.stm.

Litigation against tobacco companies. *U.S. Department of Justice Civil Division*. Updated June 8, 2001.
www.usdoj.gov/civil/cases/tobacco2.

1.4   Chlorpyrifos: A Killer With a Secret

Thousands of apples in a huge vat of chemicals in Chile.

The large-scale spraying of crops with organophosphates has been commonplace in the USA and elsewhere, yet information about the adverse effects has allegedly been concealed by manufacturers.

In 1965, the Dow Chemical Company launched an organophosphate compound, chlorpyrifos, onto the American market. It appeared under the brand names of Dursban and Lorsban and, in time, proved immensely popular as a means of pest control, reaching annual sales of over $2 billion and becoming one of the top five insecticides used worldwide. It has been estimated that Americans were spraying and shaking between 9.9 million and 13.9 million pounds (4.5 million and 6.3 million kilograms) of Dursban around their homes every year to kill cockroaches, termites, ants, fleas and mosquitoes. Up to 12.1 million pounds (5.5 million kilograms) of chlorpyrifos, under various brand names, was also being sprayed annually over crops, many of which were destined for supermarkets and the plates of millions of American families.

From 1989 to 1997, chlorpyrifos was manufactured by DowElanco, part owned by Eli Lilly and Company, and subsequently by a Dow subsidiary, Dow AgroSciences, which declared it to be of "proven safety and quality." According to a company statement, "more than 3,600 studies and reports have been conducted examining critical aspects of chlorpyrifos products as they relate to health and safety. Taken together, these reports and studies affirm that chlorpyrifos products, when used as directed, provide wide margins of safety for both adults and children."

How was it, then, that in June 2000 the U.S. Environmental Protection Agency (EPA) found chlorpyrifos sufficiently dangerous that it should be banned from use in homes, gardens and some agricultural applications, ordering that most indoor uses should be phased out over a five-year period? The unfolding story amounts to a classic case of commercial coverup.

The case so far

It has been alleged that for 10 years, from 1984 to 1994, both Dow and DowElanco concealed medical reports about adverse health effects associated with chlorpyrifos exposure. In 1997, in a case brought in a Virginia district court against various defendants, including Dow, it was revealed that Dow had been required to submit data about the testing of Dursban and its toxicity after the EPA registered the product as a pesticide in 1981. By June 1989, however, the EPA decided that the studies submitted were "wholly unacceptable" under current registration standards and stated that "data gaps on the toxicity of chlorpyrifos remained in an array of areas relating to human exposure."

The case exposed a catalog of alleged concealments of information from the EPA. It was disclosed that between 1984 and 1988, Dow had "failed timely to report nearly 100 incidents of unreasonable adverse effects from chlorpyrifos to the EPA." These incidents were not reported to the agency until late in 1994, in some cases more than 10 years after they had occurred. Yet the delay in passing on the information was in direct contravention of the U.S. Federal Insecticide, Fungicide and Rodenticide Act (FIFRA), which controls the manufacture, sale and use of insecticides. The act requires that "adverse effect incidents" must be reported within 30 days of the incident. In 1995, the EPA filed a formal complaint against DowElanco, citing 288 incidents where the company had failed to comply with the ruling.

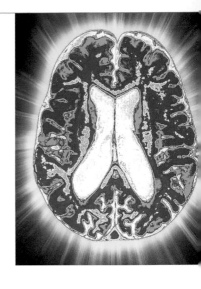

The strange glow around the image of a patient's brain in this MRI scan evidences the damage caused by organophosphate poisoning.

Researcher placing samples in an auto-sampler during toxicology research investigating the toxic effects on living tissue of organophosphate chemicals.

A woman with virtually no protection from the harmful effects sprays an organophosphate pesticide onto crops.

Liar, liar

When DowElanco eventually submitted the reports to the EPA, it became clear that chlorpyrifos might be a significant cause of chronic nervous complaints, including headaches, dizziness, mental confusion, mood swings and other symptoms. The agency review, finalized in January 1997, found that "DowElanco had incorrectly stated that it was unaware of symptoms of organophosphate poisoning having occurred with the use of chlorpyrifos." A voluntary agreement was reached to discontinue many of the uses of the chemical, including all spraying and fogging. It was also agreed that the labeling of pesticides containing chlorpyrifos was to be changed and that further research was needed into the effects of exposure to chlorpyrifos. A U.S. government study carried out in 1994 found that over 80 percent of Americans had detectable levels of trichloropyridinol (one of the main ingredients of chlorpyrifos) in their urine, suggesting that most Americans are now chronically exposed to the chemical. According to the Gardener's Supply Company, the studies found the chemical in 92 percent of adults and 89 percent of children who were tested.

REFERENCES

Donnay, A. Chlorpyrifos (a.k.a. Dursban)-the facts. Source: DowElanco, 1994. DowElanco submission to the United States Environmental Protection Agency on claims related notifications to the Agency from October 13, 1994, to November 4 1994. *Bhopal (Union Carbide Corporation)*. Undated.
www.bhopal.net/chlorpyrifos.html

Dow AgroSciences Canada Inc. announces changes in the use of chlorpyrifos products. *Canada NewsWire*. September 2000.
www.newswire.ca/releases/September2000/25/c6853.html

Dow's vital statistics. *Bhopal (Union Carbide Corporation)*. Undated.
www.bhopal.net/dowImbalance.html

EPA bans nation's top insecticide. *Gardener's Supply Company*. June 2000.
www.gardeners.com/community/Gajuneoob.asp

EPA bans pesticide Dursban, says alternatives available. *CNN*. June 8, 2000.
http://cnn.com/2000HEALTH/06/08/dursban.ban.02/

James, S. Union Carbide, Dow-Images of Bhopal, Vietnam War. *Reuters New Service*. August 15, 1999.
www.bhopal.net/UCCDowimages.html

Lescs v. Dow Chemical Co. Briefs. *Public Citizen*. Undated.
www.citizen.org/litigation/briefs/Preemption/articles.cfm

Shaw, R. Pesticide on trial with EPA. *Environmental News Network*. CNN. January 25, 2000.
www.cnn.com/2000/NATURE/01/25/pesticide.enn

Summary of the hazards of Dursban. *Natural Resources Defense Council*. September 5, 2000.
www.nrdc.org/health/pesticides/bdursban.asp

1.5   Cell Phones: How Safe Are We?

Millions of parents buy their children cell phones for safety and security, but at what cost?

By 1999, cell phone sales in the United States had topped $37 billion, and in 2000 the figures were still rising. Predictably, the makers of cell telephones and the providers of communication services insist that they are safe for us to use. Norman Sandler, Motorola's director of global strategic issues, reassured customers in 2001 that "there is absolutely no credible scientific evidence of any health risks associated with the use of wireless phones." A spokeswoman for one of the major U.S. carriers, Verizon Communications, issued an equally dismissive and soothing statement. "The available scientific evidence doesn't demonstrate any adverse health effects," she said.

Yet, as early as 1999, under pressure from medical experts, the U.K. government felt obliged to establish an independent team under the chairmanship of Sir William Stewart to study the possible adverse effects of cell phone use. In April 2000, it was revealed that safety regulators of the Food and Drug Administration (FDA) in the United States were also investigating the issue, based on two industry-sponsored studies that it was felt required additional consideration. These studies, funded by the American Cellular Telecommunications Industry Association (CTIA) to the tune of $27 million, were aimed at assuring consumers about the safety of cell phones. One of the studies, however, carried out by the American Health Foundation in New York, claimed to have detected a link between cell phone radiation and a type of brain tumor. The other, by Integrated Laboratory Systems, found that there was a breakdown of DNA in human blood cells when exposed to large doses of cell phone radiation, "possibly laying the genetic groundwork for cancer."

A computerized image of the distribution of heat following the use of a mobile phone.

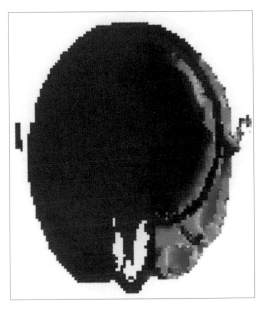

It is now believed that cancers can be triggered by a reaction to cell phone radiation. Yet the industry continues to dispute scientific claims.

In denial

The CTIA brushed the research findings aside, its vice president for external and industry relations, Jo-Anne Basile, describing them as "just pieces of a much larger picture, and the larger picture continues to reaffirm the fact that there are no adverse health effects from using cellular phones." Yet these warnings do not stand in isolation. At the University of Washington, bioengineer Dr. Henry Lai has published several research papers indicating that rats suffered measurable genetic brain damage after being exposed to a similar type and strength of radiation to that given off by cell phones. Other independent researchers at the American Health Foundation studied the incidence of a particular kind of brain tumor that grows from the edge of the brain inward (known as a neuronal tumor) in cell phone users compared with a control group. They discovered that 14 out of 34 patients with that type of cancer used cell phones, and their conclusion was that the users had more than double the risk of developing the neuronal tumor. A medical physicist, Philip Dendy, writing in the British medical journal the *Lancet*, has warned that "in the light of experience with ionizing radiation and radioactive materials, out-of-hand dismissal of the possibility of subtle effects of low-intensity, pulsed microwave radiation is most unwise."

In May 2000, the Stewart Report, published by the British working group of the same name, made various precautionary recommendations concerning cell phone use, but in December of that year the cell phone industry was accused by a leading academic at the University of Warwick, Dr. Gerald Hyland, of exploiting the gray areas in the report in order to "obfuscate the issue." Research published in the *Lancet* has also indicated that not only is the level of risk unclear, but also that cell phone operators may be using uncertainty to confuse buyers.

In June 2001, scientists in Australia added to the concerns when research was published indicating that cell phones may cause cancer at lower radiation levels than were previously recognized to be potentially

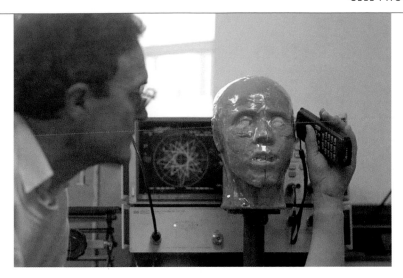

A scientist investigating the possible health risks of using cell phones runs tests on a model head, complete with artificial skin and other tissues, in order to evaluate the effects of radiation.

dangerous. The Australian findings include the possibility that certain cancers are triggered when the body is forced to produce unusual levels of heat shock proteins (also known as stress proteins as they are over-produced when the body feels extremes of heat, cold and oxygen deprivation) as a reaction to cell phone radiation.

The issue heats up

In April 2001, a number of cell phone companies were named as defendants in two U.S. lawsuits, alleging links between cell phones and possible health risks. The *Washington Post* reported that the actions named 25 of the largest companies in the industry, including such well-known names as Motorola, Nokia and Ericsson. In the lawsuits, links are alleged between cell phones and damage to basic brain function, genetic irregularities and other susceptibilities to toxins and infections. In lawsuits filed in various U.S. States, one of the lawyers acting for the plaintiffs, Peter Angelos, accused cellular telephone companies and equipment manufacturers of knowingly selling dangerous products that have inflicted radiation damage on customers.

Cell phone industry representatives continue to dispute many of these claims and argue that cell phones remain safe. Who are we to believe?

REFERENCES

Ahmed, K. Mobiles to carry health alert. *Observer Expert Group.* EMFacts Consultancy. April 30, 2000.
www.tassie.net.au/emfacts/mobiles/observer.html

Allen, P. Mobile phone firms "fudging safety issues." *Network News.* December 7, 2000.
www.vnunet.com/News/1115176

Ananova. Mobile phone firms named in lawsuit over alleged health risks. *RF Safe LLC.* April 20, 2001.
www.sartests.net/articles/ananova_042001.htm

Left, S. The mobile cancer debate heats up. *The Guardian.* June 26, 2001.
www.guardian.co.uk/Print/0%2C3858%2C4210967%2C00.html

Rosenberg, E. and J. Corbett Dooren. Cell-phone health risks need to be studied, FDA says. *(Seattle Post-Intelligencer: Washington Bureau.)* EMFacts Consultancy. April 1, 2000.
www.tassie.net.au/emfacts/mobiles/fcc_waring.html

1.6   Deep Vein Thrombosis in Economy Class

Passengers in economy class of aircraft sit in very cramped conditions. Commonly known as Economy Class Syndrome, DVT in airline passengers has been linked to the lack of room for movement in aircraft.

In July 2001, retired University of Cape Town professor of medicine Peter Beighton disclosed some astonishing information about "Economy Class Syndrome"—a popular term for a potentially fatal condition known as deep vein thrombosis (DVT), which airline passengers risk sustaining. British Airways, Qantas and KLM are facing lawsuits in Australia and the United Kingdom for failing to warn passengers about the dangers of DVT. Suggestions have been made that Air France may also be sued. If the plaintiffs are successful, the cost to the airlines may run into millions of dollars.

Initial findings

In 1967, Beighton worked as a member of the medical staff at the Hillingdon Hospital, London, which oversaw all medical emergencies from Heathrow Airport. During the following year, he and another physician, Peter Richards, working for Air Corporations Joint Medical Services at Heathrow, collaborated on a study paper analyzing the illnesses of 25 airline passengers admitted to the hospital between 1963 and 1965. The paper, said to have been made available to all airlines, warned of the risk of DVT on long-haul flights. It drew the conclusions that long periods of immobility combined with pressure from an aircraft seat against the back of the legs and the shortage of oxygen in the cabin generated "an ideal climate for precipitating deep vein thrombosis." The paper was published in the British Heart Journal and was followed up with further correspondence from Richards in 1973.

Beighton's disclosure conflicts somewhat with a statement from British Airways that it was only during the 1990s that the airline "started to see a groundswell of opinion looking at a possible link between DVT and long-haul travel...with the benefit of hindsight, it is easy to reflect on what the industry's priorities should have been." Beighton firmly refutes this statement, and the allegation that airlines have known about the risks for 30 years is also supported by the director of the U.K.-based Aviation Health Institute (AHI), Farrol Kahn.

Thirty years later

The issue of DVT was first brought to the attention of the public in the United Kingdom in November 1998, when the *Observer* newspaper uncovered a dozen cases of DVT leading to ill health, amputation or death and quoted Kahn as saying, "The U.K. airline industry is reluctant to bring this into the open." Other expert witnesses confirmed to the *Observer*'s reporter John Sweeney that the problem had been covered up and that the risks were not explained to passengers.

Claims that airline company doctors had been aware of the risks associated with immobility on long-haul flights since 1968 were also raised first by the *Observer*. This was revealed in January 2001, when it was also alleged that British Airways had not allowed doctors access to its passenger database in order to conduct further research. At that time, Kahn is quoted as saying, "There has been a major failing of the protection of public health." Kahn offered a conservative estimate that 30,000 cases of DVT during long-haul flights in the cramped conditions of economy-class seating occur each year.

In 1998, a spokesman for the Civil Aviation Authority (CAA) in the U.K. voiced the argument, "There is no medical evidence that deep vein thrombosis occurs more frequently in air travelers than in the general population." More recently, a spokesman for British Airways is quoted as saying, "There is little evidence to show that DVT is caused by flying—it could occur following any long periods of immobility." Yet this view is not shared by a growing number of medical experts. Dr. John Scurr of University College London states, "There is a volume of anecdotal evidence from hospital doctors of a relation between blood clots and long-haul flights. The CAA's attitude is that,

This simple diagram reveals the route taken by a blood clot breaking loose from a vein in the leg and travelling to the lungs.

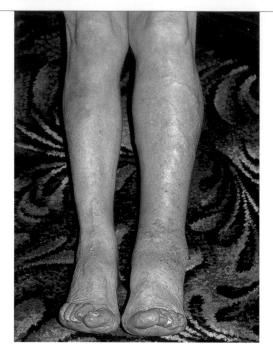

The swollen left leg of a patient after DVT has been triggered by a combination of slow-moving blood flow and an additional factor that leads to clotting. When such a clot breaks free and reaches the lungs, it may block an artery causing a serious condition known as pulmonary embolism.

if the problem develops after the flight, it's nothing to do with them." Peter Beighton perhaps more accurately sums up the position when he states, "The problem is that it is a problem of economics—the more seats and the more cramped they are together, the cheaper it is for the airlines." Were companies to recognize the existence of a danger, they would be faced with enlarging seat spaces, thereby reducing passenger and forcing up ticket prices.

Independent research carried out in 2002 by the AHI reveals that young passengers are just as likely to suffer DVT as older people. Ruth Christoffersen, chair of the Victims of Air Related DVT Association, whose daughter died of deep vein thrombosis at age 28, claims, "Airlines have little respect for human life. Passenger deaths are treated with contempt." Several companies, including British Airways and Virgin Atlantic, have now started to point out the dangers to the traveling public, but many other airlines keep silent. In January 2001, British Airways highlighted a study on its Web site that played down DVT risks on long flights.

Facing up to the truth

In April 2001, the AHI claimed that airlines were not doing enough to address the problem and could face litigation by burying their heads in the sand. The advice proved appropriate as three months later, an Australian law firm, Slater & Gordon, issued writs in the Supreme Court in Victoria on behalf of three test case claimants against British Airways, Qantas, KLM and Australia's Civil Aviation Safety Authority. A week later, a Hertfordshire law firm commenced a similar legal process in the U.K. on behalf of 30 British victims of DVT, 10 of whom had died by that time. A spokesman for the firm warned that if these initial claims met with success, the potential existed for a floodgate of writs to be lodged against various airline companies. He commented, "There are a lot of people out there who would like to bring claims."

British Airways' early response was to maintain that it does everything possible to warn passengers of potential risks, and the company announced that it wished to learn more about DVT. They were now supporting independent DVT research and "any links with all forms of transportation including flying." Whatever remedial actions airlines choose to take, however, they may amount to too little, too late.

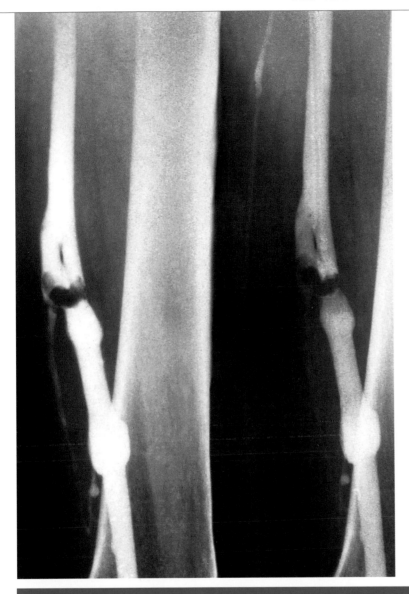

Colored X-rays of two views of a leg showing DVT. The abnormal clot (in red) is blocking the flow of blood (in white) through the vein.

REFERENCES

Airlines face DVT legal action. *BBC News Online.* July 30, 2001.
http://news.bbc.co.uk/hi/english/uk/newsid_1463000/1463757.stm

Airlines sued in DVT test cases. *Ananova.* July 16, 2001.
www.ananova.com/news/story/sm_352863.html

The cramped skies. *The Observer.* January 14, 2001.
www.observer.co.uk/focus/story/0,6903,422037,00.html

Debate continues over carriers' response to health scare. *World Airline News.* April 27, 2001.
www.findarticles.com/cf_dls/m0ZCK/17_11/73827423/pl/article.jhtml

Farrant, D. DVT victims launch suit. *The Age.* July 18, 2001.
www.theage.com.au/news/national/2001/07/18/FFX5XT6K8PC.html

Long-haul hell. *The Observer.* January 14, 2001.
www.observer.co.uk/leaders/story/0,6903,422008,00.html

Naidoo, M. Airlines "kept deadly secret for years." *Independent On Line.* July 28, 2001.
www.iol.co.za/index.php?click_id=79&art-id=qw996318782157A622

Sweeney, J. Health hazards in economy class. *The Observer.* November 22, 1998.
www.observer.co.uk/uk_news/story/0,6903,529217,00.html

Tucker, K. "Economy-class syndrome" a major health threat. *World Socialist Web Site.* January 26, 2001.
www.wsws.org/articles/2001/jan2001/dvt-j26_prn.shtml

Young "just as vulnerable to DVT." *Ananova.* May 9, 2002.
www.ananova.com/news/story/sm_584512.html

1.7   Vehicle Safety Secrets

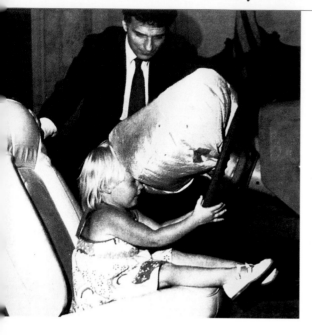

During the 1960s, Ralph Nader became the scourge of the American auto industry by publicizing car-makers' failures to make vehicles safer in ways that science had proved possible. In 1965, he published a book entitled *Unsafe at any Speed*, in which he attacked various manufacturers, including General Motors, for what he insisted was a dangerous fault in the suspension design on the Chevrolet Corvair. The book caused such a furor in the United States that the government deemed it necessary to pass the National Traffic and Motor Vehicle Safety Act in 1966. Yet to what extent has Nader's pioneering attitude resulted in better vehicle safety? Matters have improved over subsequent decades, yet all is not perfect. Evidence continues to emerge suggesting that car-makers are secretly exploiting loopholes and putting profit before safety.

May 7, 1977: Consumer advocate Ralph Nader with a young girl in a demonstration of the air bag system. She sits almost calmly as the bag 'explodes' in her face.

GM fuel tanks

In 1973, Edward C. Ivey, an engineer working in the Oldsmobile division of GM, sent an infamous memo to the senior management discussing the "cost-effectiveness" of fuel systems. Statistics showed that up to 500 victims burned to death each year in GM cars. Ivey reportedly placed a theoretical "value" of $200,000 on each fatality, which when multiplied by the number of deaths equals $100 million; he then extrapolated a cost figure of $2.40 for each of the 41 million cars on the road. The reasoning that emerged was that if GM could install a safer fuel tank for less than $2.40 it stood to save money. The actual cost to provide safer fuel tanks, however, would be $8.59 per car. Meanwhile, GM fuel tanks continued to burst and burn in road crashes. In 1981, Ivey was interviewed by a GM defense attorney about his findings; the legal representative subsequently wrote, "Obviously Ivey is not an individual whom we would ever, in any situation, want to be identified to (plaintiffs) and the documents he generated are undoubtedly some of the most potentially harmful and most damaging were they ever to be produced."

The 1963 rear-engine Chevrolet Corvair Monza GT, which could reach speeds of 186 miles per hour (300 kilometers per hour). In 1965, Ralph Nader said that there was a dangerous fault in its suspension.

General Motors, it is alleged, hid the Ivey interview memo until 1998, when a Florida state judge ordered it into evidence in a case. In 1991, two children had burned to death in an Oldsmobile station wagon; the car had caught fire following a rear-end collision that had ruptured its fuel tank. It was claimed that GM had "repeatedly tried—and continues—to quash motions requesting the evidence and to seek protective orders based on claims that the documents are protected by attorney-client privilege." In 1999, the same memo was produced in a California lawsuit against GM relating to a Chevrolet Malibu crash in 1993. On that occasion, the jury awarded $107 million in actual damages and $4.8 billion in punitive damages (later reduced by the judge to $1.09 billion). The jury concluded that the fuel tanks on the car were too close to the rear bumper and that GM had rejected a "$9 per car fix," while knowing the dangers involved with the design.

Jeep Grand Cherokee

The GM case does not stand alone. DaimlerChrysler AG manufactures an extremely popular sport utility vehicle (SUV) known as the Jeep Grand Cherokee. About 1.3 million were sold between 1995 and 1999, and in the first six months of 2001 alone, the company reportedly sold over 100,000 in the USA. In July 2001, however, the National Highway Traffic Safety Administration (NHTSA) launched an investigation after receiving complaints that the vehicle could go into reverse without warning, while idling in Park. These incidents were first brought to light in the *Los Angeles Times* for having apparently caused 32 crashes and 14 injuries. What was not revealed to the newspaper at the time were three deaths that had reportedly occurred when Jeep Grand Cherokees trapped and crushed bystanders and motorists, having unexpectedly lurched into reverse. At the beginning of August 2001, the *Los Angeles Times* ran a follow-up article disclosing that DaimlerChrysler had settled at least four lawsuits arising from accidents, one of which involved a fatality. DaimlerChrysler did not acknowledge liability in the settlements and initially claimed that the problems were likely attributable to driver error.

The NHTSA had, by that stage, widened its investigation after receiving reports of a death allegedly linked to the Grand Cherokee's potential safety defect and of confidential settlements with plaintiffs. The *Los Angeles Times* said it had confirmed two other fatalities through interviews, police reports and court records. It was also reported that the NHTSA had received 144 complaints of "inadvertent rollaway in reverse," including about 100 accidents and 40 injuries. Meanwhile, the NHTSA investigation continues. By December 2001, the number of complaints had reached 865, including 359 crashes, 184 injuries and five deaths. It is said that the NHTSA had also managed to duplicate the error during

The Jeep Grand Cherokee, which allegedly could go into reverse without warning while in Park, was the subject of lawsuits in which DaimlerChrysler initially claimed that accidents were probably due to driver error.

↑

A Bridgestone/Firestone lab technician prepares samples from recalled tires for various physical tests.

testing, yet DaimlerChrysler allegedly maintained that it had not found "any evidence to explain the sudden gearshifts."

Making progress?

In 2000, another dispute involved Firestone tires and the Ford Motor Company, who wrangled over legal liability relating to fatal accidents allegedly linked to tires on Ford SUV vehicles. Such examples of reported corporate secrecy in the U.S. automobile industry have brought public outrage, leading U.S. Congress to pass a bill requiring manufacturers to provide early warning of potential safety defects. The Transportation Recall Efficiency, Accountability and Documentation (TREAD) Act, 2000, has required the NHTSA to establish safety rules that address the proper reporting of defects by June 1, 2002. But critics have argued that the legislation may do little to protect consumers from defective automobile products. It has been pointed out by Joan Claybrook, a former head of NHTSA and now president of a consumer watchdog association, that by not keeping records of some problems, manufacturers are not bound to report that information to investigators. Car-makers can also delay matters, Claybrook says, by withholding information until they are "about to be caught," since, under the terms of the legislation, they can avoid penalties by coming forward in a reasonable amount of time and revealing information that was previously withheld.

REFERENCES

Crash and burn: GM adding fuel to fires. *Safetyforum.com.* Undated.
www.safetyforum.com/gmft/

Feds widen Jeep Cherokee investigation. *Claris Law, Inc.*
December 11, 2001.
www.injuryboard.com

Firestone tyre recall. *American Legal Network.* Copyright 2001.
www.americanlegalnetwork.com/defective_products/

Firestone tires. *Claris Law, Inc.* Copyright 2002.
www.injuryboard.com/lvlThree Cause.cfm

Grand Cherokee probed. *CNN Money.* July 4, 2001.
http://money.cnn.com/2001/07/04/home_auto/cherokee/

Jeep safety probe widens. *CNN Money.* August 3, 2001.
http://money.cnn.com/2001/08/03/news/jeep_cherokee/

Skertic, M. Critics assail auto-safety bill. *Chicago Sun-Times.*
October 30, 2000.
www.suntimes.com/tires/october/tyre30.html

St. Pierre, N. The Firestone fiasco: Was the NHTSA "asleep at the wheel?" *Business Week Online.* September 8, 2000.
www.businessweek.com/bwdaily/dnflash/sep2000/nf2000098_774.htm

2 Doctor Knows Best

Medical ethics under the knife

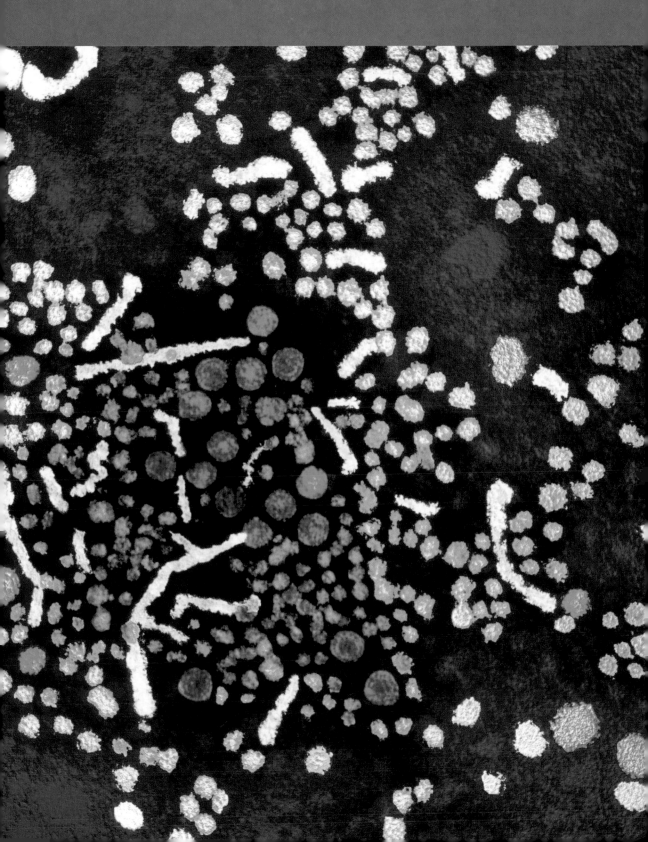

2 Doctor Knows Best

Despite all the good intentions and social benefits that medical progress holds, things are not always as wonderful as they seem. Pharmaceutical companies and research organizations are more often than not commercial enterprises in their own right, or are funded by large corporations, with budgets to be met and money to be made.

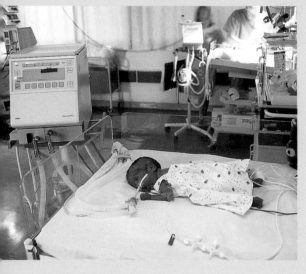

A premature baby in a hospital incubator. Thousands of people catch bugs in hospitals in the West, but thanks to the liberal handing out of antibiotics in the past, they are not always effective at treating the conditions of today.

More often than not, the long-term objective of such research programs is to cover the costs of the development of the products or theories—which can run into millions of U.S. dollars—and turn them into commercially viable end products. In this respect, medical progress is no different from any other business activity; trade secrecy is a major factor in financial success. This, in itself, is not necessarily a bad thing. But questions must be asked when those involved find a product ineffective or harmful, after millions of dollars have already been spent, yet they keep the information concealed from the public and regulatory bodies.

Researchers and manufacturers, however, are not always the only ones to blame. Physicians all over the world prescribe unnecessary antibiotics under pressure from their patients, and over the course of time, this has created drug-resistant strains of common bacterial infections. In other circumstances, when doctors make mistakes, their professional pride and concern over their careers being cut short can greatly influence a decision of admittance or false innocence. Politicians and government agencies make decisions on behalf of the general public, for example, on the basis of scientific research. But what if that research is proven to be unfounded? Will governments admit errors of judgment have been made and accept the financial consequences of compensation to those affected, or will they maintain, in public at least, that their research is correct? In cases still awaiting conclusive answers, only time will tell.

In some instances, particularly where belated examination reveals medical practice to have gone seriously awry, the problems can be traced to a gradual lessening of standards. By the time errors become obvious to other members of the hospital or medical research staff, there may already be a coverup of scandalous proportions. It then becomes not so much a moral dilemma, but a matter of self-preservation whether to come clean and face the inevitable fallout or keep quiet and hope that the situation disappears. Rarely, of course, does this happen, and when the evidence of failure and questionable practice is swept under the carpet for as long as possible, matters are invariably made a great deal worse. Such would appear to have been true in the infamous saga of infant heart deaths that occurred at the Bristol Royal Infirmary, in the United Kingdom, during the late 1980s and early 1990s. The inquiry that followed lifted the lid on what amounted to an old-boys' culture among doctors, involving secrecy about performance and lack of external monitoring. In such cases, jealously guarded professional reputations are usually at stake. If someone is seen to have bungled a research program or a medical procedure, future career prospects can take a rapid downward turn.

The upside of medical secrets being revealed, though, is that the recommendations from investigations can apply to other instances and prevent future potential errors or coverups from taking place. In the Bristol Royal Infirmary scandal, for example, government reforms of hospital procedure followed the investigation, including the independent monitoring of standards and more openness about clinical performance. However, the problem is that what applies in one country does not necessarily apply in another; regulations about medical practice and the sale of pharmaceuticals vary from nation to nation. Only when detrimental products and methods are prevented in all countries to all people will justice be served.

The Thalidomide tragedy highlighted the tremendous responsibility that pharmaceutical companies hold. It led to the deepening of medical ethics and the development of more in-depth pharmacological research today.

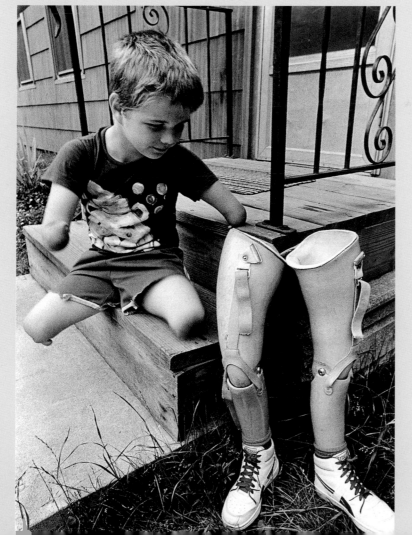

2.1   The Thalidomide Tragedy

May 22, 1968: Six of the executives of the pharmaceutical company that produced Thalidomide who were due to go on trial accused of manslaughter and causing deformity, sickness and bodily harm through the sale of their drug.

November 19, 1962: Richard Satherley, whose mother took Thalidomide during pregnancy, uses his toes to knock out a tune on his toy piano.

The case history of the drug thalidomide, which caused an estimated minimum of 8,000 babies worldwide to be born with limb defects in the 1950s, is included in this book not necessarily for the reasons that most might suspect. Despite claims that the drug companies marketing thalidomide were "dishonest and ruthless" in their behavior, there is no evidence that they either knew of or guessed the grave potential problems associated with the drug. If a coverup exists, it is one that is currently in the making.

The story so far

Thalidomide was first synthesized in Switzerland by Ciba in 1954, but they discontinued its development having found no viable therapeutic benefits. Nonetheless, the drug went into manufacture three years later at a small German pharmaceutical company, Chemie Gruenenthal, to alleviate the symptoms of morning sickness and nausea associated with pregnancy. Subsequently, several other companies, including the Distillers Company in the U.K., prepared the drug and it was marketed extensively around the world under brand names such as Contergan, Distaval and Kevadon, with firm assurances that it was nontoxic, with no side effects and "completely safe for pregnant women."

In 1960, Chemie Gruenenthal applied for approval to market the drug in the United States, but the request was rejected on the grounds that the research on which the application was based was thought to be incomplete and inadequate. The licensing application was resubmitted in January 1961, but by then the first alarm bells were ringing in Europe. It was being claimed that a toxic effect known as peripheral neuritis had been detected in patients taking thalidomide. Furthermore, Australian obstetrician William McBride had pinpointed thalidomide as the source of an epidemic of babies born without fully developed arms and legs. Another physician, Widukind Lenz, investigating outbreaks in Germany and Canada, had reached similar conclusions.

The U.K. authorities took the drug off the market, though not until November 1961. Germany, however, dismissed the findings for some time, only initially agreeing to change the status of the drug from "over the counter" to '"prescription only." The drug was never sold legally in the U.S. Frances Kelsey, a scientist appointed by the U.S. Food and Drug Administration (FDA), had been concerned over the lack of significant data and pressed for a delay in the U.S. approval that lasted long enough for the full extent of the problems to be identified. It has been alleged that the FDA was kept ignorant of the developing connection between thalidomide and babies born with limb defects, but there is no firm evidence to support this accusation. Eventually, the catastrophic side effects of the drug became exposed to the full light of publicity on both sides of the Atlantic. Letters began to appear in such prestigious publications as the British medical journal the *Lancet*. In January 1962, Lenz wrote, "I have seen 52 malformed infants whose mothers had taken Contergan...I have received letters...reporting 115 additional cases."

Accusations were leveled against several of the pharmaceutical companies, alleging that they had not conducted proper research on thalidomide, including tests on pregnant animals. Legal claims were mounted and settlements were awarded. Yet it seems clear that adequate testing was not deliberately avoided in order to market an unproven product, rather that in the 1950s no one guessed that the drug would adversely affect an unborn fetus. The kind of procedures that are now mandatory in many countries to ensure drug safety were simply not required then, and it was only cautiousness on the part of the FDA that delayed approval in the U.S.

Through the generations

If there is a coverup to be investigated, it is one that is happening today. In 1994, William McBride and a medical colleague in Australia, Peter Huang, published research carried out on laboratory animals indicating that the effects of thalidomide could be passed from one generation to the next. In 1997, the same team published further research pointing more strongly to the possibility that thalidomide bonds to DNA (the molecular blueprint of life) and that it interferes with cell replication. In the U.K., the

Richard Satherley, a few years later, with the arms that mean so much to him. His new limbs were powered by a tiny cylinder of compressed gas.

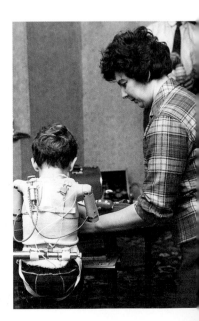

United Distillers, now owned by Guiness, were the original U.K. distributors of thalidomide. Georgina Harrison walks unaided on her artificial limbs after delivering a dossier to the brewery giant, claiming that the deforming effects of the drug can be inherited.

Sunday Mirror newspaper then reported that six British men who had been born with deformities because of thalidomide had fathered babies that had also been born with deformities. Further, in two cases the deformities more or less exactly replicate those of the parents. William McBride alleges that similar second-generation victims are alive in Germany, Japan and Bolivia.

Glenn Harrison, vice chairman of the Thalidomide Action Group (TAG) in the U.K., revealed that his daughter has disabilities similar to his own. He also reported that other parents in similar situations have been reluctant to advertise their plight because they fear that the Thalidomide Trust, which handles compensation claims in the U.K., might argue that the parents were affected by some other genetic condition and would therefore reject claims. TAG asserts that this has already happened in Germany. In 1997, Harrison commented, "The [U.K.] Department of Health should continue the research which they have been denying us for three years. We have the living proof that it could happen and the research to back it up." The Department of Health, however, maintains that there is insufficient evidence to support the theory of second-generation transfer and the director of the Thalidomide Trust has commented, "It is a matter for the trustees' discretion whether they consider a person to be affected and they [the Department of Health] have upheld the trustees' view in every case."

In the autumn of 1997, the FDA began to reconsider whether thalidomide should be approved in the U.S.—not for pregnant women but in the treatment of leprosy and for AIDS patients who were buying the drug on the black market, having found it to be therapeutic. The FDA then approved the use of thalidomide in the treatment of leprosy in July 1998, albeit with stringent restrictions on distribution. The drug is also being used again in the U.K. to treat a variety of conditions.

REFERENCES

Hirsch, J., A. Caranna, and M. Agramont. The history of Thalidomide (class project). *Richland College.* March 2000.
www.rlc.dcccd.edu/MATHSCI/reynolds/thalidomide/history/history.html

Kurata, M. A. Thalidomide's history and its consequences (unpublished student research paper). *Miye Alice Kurata Web Page.* Undated.
www.cems.alfred.edu/students/kuratama/HistoryHW5.html

Lesso, J. The thalidomide tragedy. *Campaign Against Fraudulent Medical Research.* Undated article.
www.moreinfo.com.au/avs/newveg/thalidom.html

Levine, J. Thalidomide's comeback sparks emotional debate. *CNN Interactive.* September 4, 1997.
www.cnn.com/HEALTH/9709/04/nfm.thalidomide/

Marion, J. F. Thalidomide and IBD. *Mount Sinai School of Medicine.* CCFA. August 28, 1998.
www.ccfa.org/weekly/wkly828.htm

Thalidomide returns to the UK. *BBC News Online.* October 28, 1998.
http://news.bbc.co.uk/hi/english/health/newsid_202000/202713/stm

Thalidomide shock discovery. *Disability Now.* June 1997.
www.disabilitynow.org.uk/search/97_06_Je/pg1shock.htm

Wertelecki, W. Letters to "Lancet" that made history. *IBIS.* Last updated January 21, 2002.
www.ibis-birthdefects.org/start/letters.htm

2.2 Dow Corning Breast Implants

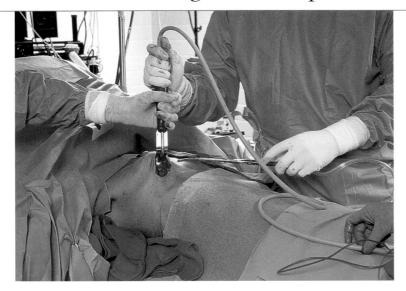

A surgeon in New York fills a patient's breast with a liquid silicone implant while safety data was allegedly withheld by the manufacturer, Dow Corning.

By conservative estimates, up to 1.5 million women around the world are currently wearing breast implants prepared from a variety of substances. Among these is silicone gel, which until March 1992 was the favored material of Dow Corning, a subsidiary of the massive Dow Chemical and Corning Corporation. Evidence shows that the gel has a tendency to leak into surrounding tissues and that this leakage can cause serious diseases.

When concern began to arise over the long-term effects of silicone implants, women in the United States started to file claims for damages against Dow Corning on the grounds that the implants affected their health and, in certain cases, led to permanent disablement. In 1988, against this background of alarm and litigation, the U.S. Food and Drug Administration (FDA) requested that Dow Corning identify all studies in which any form of silicone was injected into laboratory test animals. Dow Corning filed a report claiming that there was no link between breast implants and disease, but which, it is alleged, excluded certain studies that the company had conducted.

The Public Citizen's Health Research Group argued that Dow Corning had filed a false report on the safety of silicone breast implants and had not disclosed studies demonstrating that autoimmune and respiratory problems could result from the use of liquid silicone. In a letter to the FDA, Dr. Sidney Wolfe, director of the group, stated, "We believe that Dow Corning violated federal law by deliberately withholding safety data on its silicone used in medical devices, especially breast implants...because of the potential harm of serious illness facing tens if not hundreds of thousands of women who have received silicone breast implants since the time of filing the false report, we believe that the FDA should seek criminal prosecution of the company and responsible Dow Corning officials."

Former Director of Corporate Communications for Dow Corning Corporation John Swanson addresses a Washington press conference to unveil a national ad campaign exposing Dow's coverup about silicone breast implants.

It has been alleged that Dow Chemical developed pesticides from components of the silicone gel used in cosmetic surgery. Here a surgeon removes the material from a breast implant.

Decades of lies

In 1994, Richard Alexander, a civil trial lawyer specializing in such claims, went further by stating that Dow Chemical knew as early as the 1950s that silicone and silica, both used in Dow Corning's implants, were "bio-reactive, immunogenic, toxic and inflammatory when introduced into the human body. The company did nothing to advise the public of these hazards or to stop the sale, despite the fact that Dow Chemical had the right to control... the quality of the products manufactured and sold by Dow Corning... Clearly Dow Chemical and Corning Corporation will argue their own analysis of this evidence. Notwithstanding what they may claim, the evidence is indeed compelling. It is difficult to escape the conclusion that this violation of public trust should not go unpunished and that Dow Chemical should be held liable for the harm it has caused and most assuredly could have prevented."

Alexander's California law firm presented evidence that from 1943, Dow Chemical agreed to provide Dow Corning with its expertise in silicone compounds. It also agreed to share all newly discovered technical information until 1958. On behalf of Dow Corning, the parent company specifically undertook responsibility for researching the biological effects of silicone compounds and continued to do so until as late as 1992. Dow Corning allegedly relied on Dow Chemical to such an extent that it did not perform any studies on breast implants before the first implant was actually inserted into a woman's body. One extraordinary piece of evidence suggests that Dow Chemical entered into a secret agreement in the 1960s to develop pesticides and insecticides using components of the silicone gel that ultimately formed the substance of Dow Corning's breast implants.

Avoiding liability

Given this link between Dow Corning and its parent Dow Chemical, plaintiffs were outraged when Dow Corning filed for bankruptcy in 1995. The judge presiding over the federal bankruptcy court in Bay City, Michigan, agreed to distance the parent company from liability, making it impossible for Dow Chemical to be sued. Dow Corning then made a series of cash offers to the victims and in 1998 announced that, in principle, it had reached an out-of-court agreement amounting to $4.25 billion with some of the plaintiff's lawyers. Others, however, were less than happy. At least one expert on health hazards, Johns Hopkins University Professor of Surgery Norman Anderson, described the plan as being seriously flawed and calculated

This photograph reveals the damaging result of a silicone implant after the material has hardened causing the breast to become rigid.

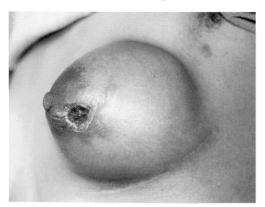

that realistic settlement of the medical damages could run as high as $169 billion. Anderson commented that the Dow Corning proposal amounted to little more than "sanctioning evasion of corporate responsibility."

Although leading lawyer for the plaintiffs Kenneth Eckstein told CNN in November 1999 that he was optimistic payments could begin in 2000, the settlement issue was still far from being resolved at the time of writing. The U.S. Appeals Court denied a motion to hear challenges by foreign claimants to the final Dow Corning bankruptcy settlement plan and sent the case back to the U.S. federal district court. The words of one victim, in July 2001, sum up the frustration: "I would like to know what is going on with the Dow Corning litigation. I am from Yukon Territory, Canada, my implants ruptured and I would really like some answers. Whenever I call my lawyer I am told to check the Web site and there is never anything new on that. Can someone please help me?"

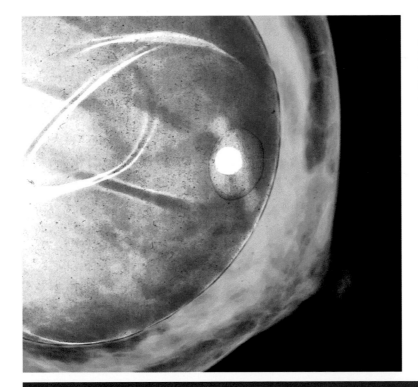

An X-ray reveals an intact breast implant. Medical evidence shows that the gel has a tendency to leak into surrounding tissues with the potential to cause serious diseases.

REFERENCES

Alexander, R. Update on breast implants: The new evidence against Dow Chemical. *Alexander, Hawes and Audet, LLP.* June 14, 1995.
http://consumerlawpage.com/article/dow.shtml

Brockley, R. Dow Corning: Breast implants gone bad. *Bhopal (Union Carbide Corporation).* Undated.
www.bhopal.net/dowcorning.html

Current Events. *The Tort Claimants Committee.* February 12, 2002.
www.tortcomm.org/tort_main.html

Garsten, E. Bankruptcy judge approves breast implant settlement. *CNN.* November 30, 1999.
www.cnn.com/US/9911/30/implant.settlement

Joint plan of reorganization and Dow Corning bankruptcy proceedings. *Lieff Cabraser, Heimann & Bernstein, LLP.* 2002.
www.lieffcabraser.com/dowbankruptcy.htm

Sissell, K. Dow Corning increases breast implant settlement plan to $4.4 billion. *Chemical Week.* February 25, 1998.
www.findarticles.com/cf_dls/m3066/n7_v160/20385608/print.jhtml

2.3 Superbugs and Antibiotics

➜

It is now widely recognized that antibiotics have been over-prescribed by the medical profession, thus reducing their effectiveness in combating infection.

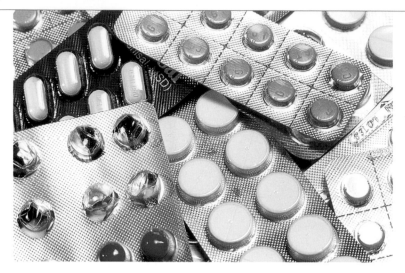

In April, 1999, the U.K.'s *Daily Mail* newspaper ran a story, "Antibacteria cleaners are creating new superbugs...instead of making the home bacteria-free, the...hygiene aids are creating a generation of potentially deadly superbugs." In September 2000, it picked up on another aspect of the same problem: "A nine-day-old baby has died from the superbug sweeping Britain's hospitals...[it] caught the antibiotic-resistant MRSA (methicillin-resistant *staphylococcus aureus*) in a special care baby unit. The infection kills 5,000 hospital patients a year."

In the millennium year, the director general of the World Health Organization (WHO), Gro Harlem Brundtland, made a more far-reaching statement concerning the microbes that threaten to defeat modern medical science: "It is a deepening and complex problem accelerated by the overuse of antibiotics in developed nations and the paradoxical underuse of quality antimicrobials in developing nations owing to poverty and a resultant dearth of effective health care." These sound bites touch on a subject that concerns us all.

Alexander Fleming, the first person to identify an antibiotic, developed penicillin in 1928.

The so-called cure-all

From the early 1940s, the medical profession and the public, who saw an end in sight to many diseases that had hitherto proved fatal, welcomed the new generation of antibiotic "wonder drug" For half a century, antibiotics have been prescribed freely, often more or less as "convenience drugs" and with little thought for the consequences. Current estimates indicate that some 51 million pounds (23 million kilograms) of antibiotics are supplied each year in the U.S. alone. Hospitals have been at the forefront of the liberal application, while family doctors and medical centers have been quick to dish out antibiotic prescriptions, often against wholly inappropriate viral forms of infection. Doctors, under severe pressure and facing unidentified organisms and conditions, have frequently prescribed antibiotics as a first resort for anything from a rash to toothache. In many countries, including the U.S., it is possible to obtain antibiotics without prescription. For a long time the public was unknowingly misled into a belief that they could obtain an antibiotic remedy for viral infections such as the flu and the common cold. They were also left ignorant of the long-term damage that could be caused by failure to complete courses of antibiotic treatment once symptoms had been alleviated. If a person stops taking antibiotics before the course is finished, because they "feel better," some bacteria will remain in a mutated form that is then resistant to those antibiotics.

The medical profession did little to educate their patients since it was easier to send them away from the office or medical center with an antibiotic prescription. Stuart Levy, a scientist at Tufts University School of Medicine and president of the Alliance for the Prudent Use of Antibiotics, confirms that doctors have been willing accomplices. He recalls that in a recent seminar, 80 percent of the physicians admitted to writing antibiotic prescriptions against their better judgment. Another physician, Eddie Hedrick, infection control manager at the University of Missouri Health Care Center, details the pressures when describing the situation of a mother bringing a two-year old child to his pediatric clinic with a simple chest infection. After several sleepless nights, she distrusted his judgment and more or less demanded the prescription of an antibiotic. Hedrick's question was, "Should I provide her with the antibiotic? If I don't, will she go to another physician who will?" He cites another case of a 40-year-old woman with symptoms of influenza (a viral disease unaffected by antibiotics) who was prescribed three different antibiotics, completed the course of none, and who eventually died of an intestinal infection because all of her normal gut bacteria had been killed off by the drugs. In many cases where people are needlessly supplied with antibiotics the cold or cough clears up of its own accord, but people are misled into belief that the antibiotics have provided a cure. Meanwhile microbes that have not been fully eliminated during the course of an infection respond by developing their own immunity to the effect of these drugs.

Resistant germs on the rise

The microbiological community has responded in ways that are both ingenious and deadly. We are faced with the rise of the "apocalypse germs"; one of the most dangerous to have emerged in recent years is MRSA. Mutant strains became resistant to penicillin as early as 1941, only months after first use, and by 1946 the figure had grown to 14 percent. Multidrug resistance was first reported in Europe in the late 1950s. Today, more than 90 percent of *S. aureus* strains are penicillin-resistant and one in particular has developed resistance to all of the antibiotics available in hospitals, with the exception of Vancomycin, the

Egg City, California, where the chickens are given feed containing antibiotics. The use of antibiotics in farming passes onto humans who eat the eggs produced.

so-called "last resort" antibiotic. Even this is no longer wholly effective against several bacterial agents. A recent report revealed that about 150,000 people become infected with life-threatening infections each year in hospital wards, approximately 10 percent of which result in fatality.

The food we eat

Commercial pressures have encouraged people in the West to "disinfect" their homes without explaining that the effect is to make us even more vulnerable to infection when it arises. Many of the bacteria that are around each day allow us to maintain natural resistance. Take them away and our resistance falls. Many people remain unaware of another "back-door" means by which they are literally being fed with antibiotics. Resistant bacteria have been permitted to evolve through the large-scale use of antibiotics as growth promoters in the livestock industry. We may, in effect, be eating our way through large quantities of such drugs as Avoparcin, which is used to control disease and promote rapid weight gain in intensive poultry farming. Stuart Levy alleges that about 40 percent of antibiotics in the U.S. are given to animals. Europe bans the use of antibiotics used in human medicine as feed additives. Such drugs, however, can theoretically enter the food chain and exert a pronounced effect on our health and safety, yet we are scarcely informed of their use, far less the implications.

A worker at a feed manufacturing plant shows a mixture of crushed corn with protein supplements that contain antibiotics.

REFERENCES

Ephraim, R Antibiotic misuse. *Conscious Choice.* July 1999.
www.consciouschoice.com/health/antibiotics1207.html

Hedrick, E. Misuse of antibiotics. Case studies. *University of Missouri. The Virtual Health Care Team.* Revised March 27, 2002.
www.vhct.org/case899/index.shtml

2.4 Ritalin and Hyperactivity

Children who suffer from hyperactivity find it difficult to sit still, cannot concentrate for long periods and often become very frustrated.

"The APA will defend itself vigorously by presenting a mountain of scientific evidence to refute these meritless allegations and we are confident that we will prevail." Thus ran an official statement issued by the American Psychiatric Association (APA) in September 2000, adding that lawsuits filed recently in courts in Hackensack, New Jersey and San Diego, California contained charges that were "unfounded and preposterous." The charges centered on the use—or misuse—of a stimulant called methylphenidate, which is marketed by the multinational pharmaceutical giant Novartis (formerly Ciba-Geigy, prior to a merger in 1996) under the brand name Ritalin.

The APA's rebuttal came as an early response to news of allegations that defendants, including themselves, Novartis and a parent organization known by the acronym CHADD (Children and Adults with Attention-Deficit/Hyperactivity Disorder), had promoted the belief that a large number of children need to take Ritalin for conditions referred to variously as attention-deficit disorder (ADD) and attention-deficit hyperactivity disorder (AD/HD). This set of conditions is said to affect between 4 percent and 12 percent of school children in the U.S., chiefly boys, who experience problems of short attention spans, impulsive behavior and hyperactivity.

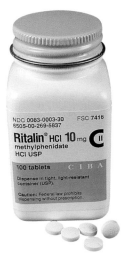

The drug Ritalin has become the subject of intense debate about its use, or misuse, in treating Attention Deficit Hyperactivity Disorder in children.

In on the act

The New Jersey and California lawsuits, on behalf of children who had taken the drug (following on the heels of a similar action filed in Texas earlier in 2000), claimed that a conspiracy had taken place among the defendants to encourage the excessive diagnosis of behavioral disorders in children and had used misleading literature in order to promote Ritalin sales. It was alleged that the conspiracy had begun in the mid-1950s and that by 1980 Novartis and Ciba-Geigy had worked

Big business. It has been claimed that in 2000, more than 9 million prescriptions for Ritalin were handed out in the U.S. alone.

successfully with the APA to have the diagnosis of ADD listed in the Diagnostic and Statistical Manual of Mental Disorders. In 1987, the listing was extended to include AD/HD, so generating, it was said, a vast potential market for the drug. It has also been claimed in the lawsuits that the manufacturers donated $748,000 to the CHADD organization between 1991 and 1994 in order to boost sales of Ritalin. It has been reported that the lawsuits were inspired by the research of a medical doctor, Peter Breggin, whose book *Talking Back to Ritalin* effectively lifted the lid on the issue and brought it to the attention of the public.

Side effects kept quiet

The suit filed in California argued that the companies failed to advise "that Ritalin usage would not stimulate or improve academic performance and/or have any long-term effect on the symptoms associated with ADD and/or ADHD." News of the legal actions has also prompted the question of whether some children are receiving unnecessary medication, and it was revealed that congressional hearings in the United States have examined whether Ritalin is being over-prescribed for the disorders. It is said that in the U.S. more than nine million prescriptions were filled for Ritalin in 2000, most of them to treat ADHD. Dick Scruggs, an attorney for some of the plaintiffs has stated, "The conspiracy is to create business for the pharmaceutical company by over-defining or loosely defining the disease such that it would fit every child in America, and creating a market for clinical psychologists to treat these kids." Another of the legal representatives involved in the Ritalin lawsuits, John Coale, commented, "They were giving this stuff away like candy."

On September 15, 2000, Novartis responded to the allegations saying that medical experts around the world have acknowledged the existence of ADHD. The evidence, according to Steven Mirin, a spokesman for the APA, goes back more than 50 years. Since then, the lawsuits filed in California and Texas have been dismissed because of insufficient proof of claim. The action in New Jersey and other actions including some in Florida and Puerto Rico are, at the time of writing, still pending.

REFERENCES

Bailey, E. About: Attention Deficit Disorder. Recent lawsuits and Ritalin. *About.com*. 2002.
http://add.about.com/library/weekly/aa051001a.htm

Breggin, Peter R. Ritalin Class Action Suits. Undated. *Peter R. Breggin Web Page*. **www.breggin.com/classactionmore.html**

Lawsuits claim Ritalin "conspiracy." *AP Health News*. September 19, 2000.
www.thirdage.com/news/ap/hlth/20000915.39c2c55e.1fbb.2.html

McKenzie, J. Conspiracy theory: Suits filed against Ritalin manufacturer, doctors. *ABC News*. September 14, 2000.
http://abcnews.go.com/onair/WorldNewsTonight/wnt000914_Rita linsuit_feature.html

Ritalin maker and American Psychiatric Association sued for conspiracy. *Chiropractic Research*. Undated.
www.chiropracticresearch.org/NEWSritalin_maker.htm

Shute, N. Pushing pills on kids? *US News (article preview)*. October 2, 2000.
http://n19.newsbank.com/

Westfield, A. Manufacturer, psychiatric group accused of overdiagnosing to sell Ritalin *Associated Press*. September 15, 2000.
http://archive.nandotimes.com/nofr.../0,2107,500257918-500396638-502363614-0,00.htm

2.5 Gulf War Syndrome

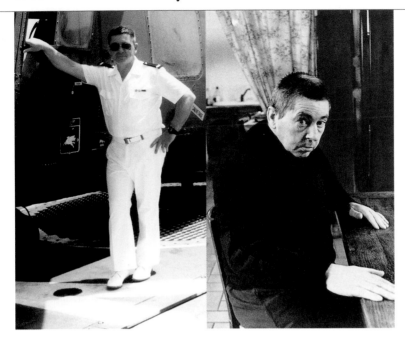

Before the Gulf War, Joseph Durrand was a fit and healthy French naval officer (left). Today (right), a shadow of his former self, he claims to be suffering from Gulf War Syndrome. Yet what is the true nature of his condition?

In many cases where allegations of secrecy and coverup arise, it is clear not only that there has been an attempt to deceive, but also that the culprits and victims can be identified with reasonable certainty. Occasionally though, the case made by the plaintiffs is unconvincing and, in the absence of proof positive to contradict the protestations of innocence by the defense, the debate remains open. One of the classic examples is that of Gulf War Syndrome (GWS). Is it fact or fantasy?

Cause and effect

Strenuous claims have been made by both American and British veterans of the war with Iraq, which ended in February 1991, that they came home suffering from an assortment of symptoms including chronic fatigue, swollen joints, digestive problems and skin rashes. Symptoms were first reported within months of the end of the Gulf War and these were attributed to an assortment of causes. Irradiation by depleted uranium used in warheads of shells has been blamed, as have the adverse effects of vaccines with which troops were inoculated against biological warfare, poisonous chemicals allegedly delivered by Saddam Hussein in Scud missiles and even the bacteria carried by a type of sand fly. In the United States, more than 70,000 personnel have complained of problems, and in the United Kingdom, a medical assessment program was set up in 1993 with which some 2,000 veterans have registered. Veterans have criticized both the U.S. Pentagon and the U.K. Ministry of Defence (MOD) for the way in which investigations into GWS were conducted and have made accusations of deliberate coverups.

British Army troops in Saudi Arabia were immunised against chemical warfare before the 1991 Gulf War.

The tip of a depleted uranium shell, of a type fired extensively during the Gulf War. Did these warheads affect the health of service personnel?

Deny all knowledge

The response of both governments has been ambivalent, conceivably suggesting collusion. In November 1996, the U.S. government maintained that investigations revealed no single coherent diagnosis, and subsequently the Pentagon testified to Congress that "there is no persuasive evidence of such exposures [to chemical and biological warfare agents] even after much scrutiny." This denial was repeated on several occasions. Yet the magazine *Covert Action* has claimed that a variety of sources, including declassified Marine Corps battlefield command chronologies and After Action Reports, concede widespread exposure to such agents when U.S.-led forces destroyed an Iraqi chemical weapons factory. It has been admitted that, in consequence, up to 20,000 soldiers may have been exposed to toxic chemicals.

Coming clean

In October 1996, having first been reticent on the subject, the British government conceded that its troops had been exposed to organophosphate pesticides during the Gulf War. In December 1996, the MOD began a new survey but allegedly declined to open files on the use of unlicensed vaccines and anti-nerve gas medication on British troops. In October 1997, however, it was conceded that the army had vaccinated troops with unlicensed and possibly dangerous drugs, despite the concerns of health officials. In new revelations, the MOD agreed that its senior officials had either ignored or never seen a warning fax issued by the Department of Health about the dangers of administering two of the vaccines simultaneously. Another area of "fudged" response seems to have emerged in 1999 when the Pentagon accepted that an anti-nerve gas pill given to troops could not be ruled out as a cause of illness. The MOD poured cold water on this, responding that the report did not support claims of the drug causing GWS.

A blurred line

Some of the voices of protest that GWS exists and that there has been an unprincipled coverup reveal a distinctly ideological tone. In May 1997, a New York-based activist organization, the International Action Center, published a book consisting of a set of essays on the use of depleted uranium (DU). Its Web site, publicizing the book Metal of Dishonor, uses particularly emotive phraseology when it describes "how the Pentagon radiates soldiers and civilians with DU weapons: the contamination of the planet by the United States military. In addition to exposing the deadly duplicity of the Department of Defense, the book documents the genocide of Native Americans and Iraqis by military radiation, the connection between depleted uranium and Gulf War Syndrome."

Even official statistics on GWS are liable to cause confusion and have been disputed among various organizations and individuals. A MOD report into GWS concluded that Gulf War veterans were apparently twice as likely to be ill as service personnel who did not serve in the Gulf.

Yet, in November 1996, the *New England Journal of Medicine* published a report on two separate studies carried out in the U.S. that cast doubt on the actual existence of the problem. According to the studies, contrary to MOD findings, the incidence of symptoms among Gulf War veterans is not significantly different from that of military personnel who did not go to the Gulf. A team at the Naval Health Research Center in San Diego examined hospital records for more than half a million Gulf War veterans against those of a control group and found only slightly higher hospitalizations among the Gulf War sample, attributed to stress and to the tendency to wait until after the war to seek treatment for medical problems. A study for the Department of Veterans Affairs came up with similar results, with the Gulf War veterans generating slightly higher figures, attributed mainly to accidents rather than disease.

It has been suggested by at least one independent specialist, Arthur Kerschen, a molecular and cellular biologist with an American-based biotechnology company, that GWS is probably attributable to a combination of factors, including the media hype surrounding the possibility of chemical and biological weapons attack during the Gulf War, the litigious nature of American society and the fact that, as a target for litigation, the U.S. government possesses extremely deep pockets.

 A United States soldier conducts a biological decontamination exercise in the Saudi desert during the Gulf War.

REFERENCES

Bernstein, D. Gulf War Syndrome covered up. *Covert Action*. Undated.
http://mediafilter.org/caq/Caq53.gws.html

Depleted uranium. *Compiled by the Depleted Uranium Education Project. International Action Center (NY).*
www.iacenter.org/depleted/mettoc.htm

The Gulf War Syndrome: fact or fantasy? *BBC News Online*.
January 16, 1998.
http://news6.thdo.bbc.co.uk/hi/english/special_report/1998/gulf_war_syndrom.../48037.st

Gulf War Syndrome hits land mine. *Academic Press*. November 13, 1996.
www.apnet.com/inscight/11131996/grapha.htm

Kerschen, A. Gulf War Syndrome. *Undated Web page.*
http://biofact.com/gulf/

MOD cool on Gulf syndrome report. *BBC News Online*. October 20, 1999.
http://news.bbc.co.uk/hi/english/uk/newsid_480000/480263.stm

Zimmerman, D. and K. Goldstein. Head in the sand. *Salon*.
February 1997.
www.salon.com/feb97/news/news970221.html

2.6 Fluoride: Are We Being Conned?

A fluoride implant is fitted into a child's tooth. Yet classified documents in the U.S. are said to reveal evidence of the adverse health effects of fluoride.

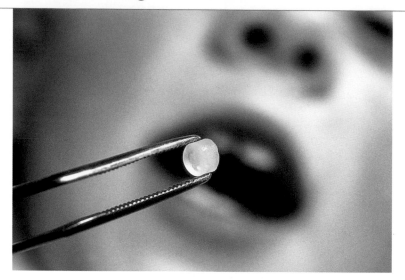

In April 1997, two American researchers, Joel Griffiths and Chris Bryson, working on a commission for the Christian Science Monitor, submitted the results of a detailed and well-documented investigation about circumstances surrounding fluoridation. Their study revealed hitherto classified information and included indications that the U.S. Atomic Energy Commission (AEC) had suppressed evidence during the 1940s concerning the adverse health effects of fluoride in public water supplies. The Christian Science Monitor allegedly declined to publish the report, but it appeared subsequently in various publications, including Waste Not. What it reveals will probably dismay anyone who has accepted fluoridation of drinking water and toothpaste as a fact of life.

There's something in the water

Some 50 years ago, fluoride was first added to public water supplies in the United States, apparently to reduce decay in children's teeth. Today, roughly two-thirds of U.S. public drinking water is fluoridated, whether American citizens wish to ingest fluoride or not. In the United Kingdom, in the autumn of 1999, it was disclosed that the British government intended to press ahead with fluoridation of public drinking water supplies despite resistance by many councils and rejection by large sections of the public. Water companies would be forced to fluoridate by law. A similar situation exists in many developed countries of the world.

Few people may realize that from the 1940s onward vast quantities of compounds based on fluorine, one of the most active and toxic chemical elements known to science, were used by the Manhattan Project in the production of weapons-grade uranium and plutonium. Nor is it generally known that the first litigations against the U.S. atomic bomb program concerned the adverse effect of fluorides, not irradiation, and that Manhattan Project scientists were covertly ordered to provide "evidence useful in litigation" against defense contractors being sued over fluoridation injury claims.

Much of the drinking water in the United States now contains fluoride. One part fluoride per 1 million parts water can reduce tooth decay by up to 65%.

Program F

Griffiths and Bryson obtained the classified and hitherto undisclosed version of a 1945 study conducted by AEC scientists with the code name "Program F." Carried out in Newburgh, New York, the program involved fluoridation of water supplies over a 10-year period from May 1945 and involved discreet collection and analysis of blood and tissue samples from residents. These samples were obtained with the cooperation of state health department staff and passed to Program F scientists at the University of Rochester. A version of their findings was then published in the Journal of the American Dental Association in 1956, asserting that "small concentrations" of fluoride were safe for U.S. citizens.

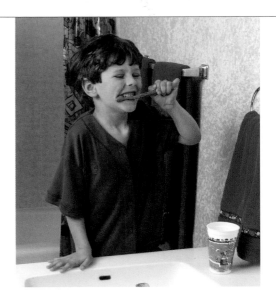

According to Griffiths and Bryson, however, while this sanitized version emphasized dental benefits of fluorides, it omitted, on the grounds of national security, the evidence of adverse health effects that had been included in the original classified version. A mayor of Newburgh, Audrey Carey, who as a child in the early 1950s was examined by doctors from the fluoridation project, has said, "It is appalling to do any kind of experimentation and study without peoples' knowledge and permission." When challenged on the ethics of secretly studying U.S. citizens to obtain information useful in litigation against the atomic weapons program, a spokesperson for the University of Rochester Medical Center commented, "That's a question we cannot answer." A Department of Energy (DOE) representative is quoted as saying, "Nothing that we have suggests that the DOE or predecessor agencies, especially the Manhattan Project, authorized fluoride experiments to be performed on children in the 1940s."

Most brands of toothpaste now contain fluoride, but independent studies in China indicate that the chemical can act as a poison on the central nervous system.

The damage has been done

At the start of the twenty-first century, the adverse effects of fluoridation are beginning to show in the teeth of American young people. It is claimed by the U.S. National Research Council (NRC) that in some large urban concentrations, up to 80 percent of the population has dental fluorosis, revealed as whitish marks on the front teeth and, in more advanced conditions, dark blotches or stripes. According to the NRC these indications point to excessive fluoride intake.

Concern over ongoing secrecy about fluoride was heightened in 1995 when Phyllis Mullenix, the former head of toxicology at the Forsyth Dental Center, Boston, Massachusetts, published the conclusions of a study on laboratory animals that had been exposed to the chemical. The research indicated that fluoride acts as a poison on the central nervous system and may affect the functioning of the human brain even at low

The intricate patterns made by fluoride crystals are revealed under a microscope. Yet their beauty may be deceptive, concealing a danger to our health.

doses. Independent studies in China appear to support her findings, revealing a link between exposure to fluoride at low concentration and diminished IQ in children. Yet when Mullenix applied for a government grant to continue her research, the application was turned down with the explanation that "fluoride does not have central nervous system effects."

However, disturbing evidence is said to exist of a coverup by the U.S. National Institutes of Health (NIH). A classified Manhattan Project memo, dated April 29, 1944, marked "Secret" and addressed to the head of the medical section, Colonel Stafford Warren, offers an altogether different picture. It includes the partly coded comments that, "Clinical evidence suggests that uranium hexafluoride may have a rather marked central nervous system effect...it seems most likely that the F [fluoride] component rather than the T [uranium] is the causative factor." Warren allegedly approved a research program into the effects of fluoride on the central nervous system based on the detailed proposal believed to have accompanied the April 1944 memo. Yet today the proposal is missing from the U.S. National Archives and may remain classified. When searching the files, the researchers also failed to locate any of the Manhattan Project fluoride CNS research papers.

The question left hanging is whether the U.S. government suppressed the findings of fluoridation studies in Newburgh and elsewhere. It is pointed out by Griffiths and Bryson that all AEC-funded studies had to be declassified before publication in civilian medical and dental journals. Where, they ask, are the original classified documents? Meanwhile, most of us living in compulsorily fluoridated areas continue to take the compound into our bodies. For how long will we be deprived of the real truth?

REFERENCES

Griffiths, J. and C. Bryson. Fluoride and the A-bomb program. *Extract from article published in Waste Not #414 (September 1997) and in Nexus Magazine Vol. 5, #3 (April-May 1998).* **www.healthy-communications.com/fluoridetoxicbomb.html.**

Griffiths, J. and C. Bryson. Fluoride, teeth and the atomic bomb. *Extract from article published in Waste Not #414 (September 1997) and in Nexus Magazine Vol. 5, #3 (April-May 1998).* **www.fluoridealert.org/WN-414.htm.**

Water fluoridation, our health and our environment. *Brent and Harrow Green Party. October 23, 1999.* **www.london.greenparty.org.uk/brentandharrow/pr_fluoridation.htm.**

A Question of National Security

Military secrecy has been a fact of life — and death — since the days of the Trojan horse. In military science, however, secrecy has come more strongly into play in the last hundred years or so.

⬆

The ongoing research and development into weapons of mass destruction will undoubtedly create arms in the future that are far more terrifying than these artillery shells. But is the defense of national security a reasonable excuse for unethical research practices and trigger-happy wars where the victims never learn what they were attacked with?

Undercover operations became a critical strategic factor during World War II when German ingenuity and inventiveness led to the creation of formidable weapons, including a swept-wing jet fighter, the V-1 cruise missile, the V-2 ballistic missile, and chemical and biological weapons. Much of Germany's rocket development was being conducted amid great secrecy on the Baltic island of Peenemunde, and it was only through some brilliant analysis of intelligence reports that the Allies were able to prevent greater disaster being inflicted on London. At the end of the war in Europe, many of the German scientists were rapidly and covertly moved to the United States in an operation code-named Paperclip. The cover story was that their efforts could be useful in shortening the war with Japan. But in reality, Operation Paperclip was mounted to prevent the knowledge these scientists had learned from falling into the hands of the Russians.

Of course, all governments need to maintain a degree of secrecy in the interests of national security, but at what point does this become morally undesirable? It can be argued that secrecy beneficial to one party may be detrimental to another—one person's meat is another's poison. In the western world, we may strongly object to the secret development of weapons of mass destruction by the regime of Saddam Hussein in Iraq and his obstructing of the UN weapons inspection teams. We may feel revulsion in the testing of those weapons on defenseless minorities like the Iraqi Kurds. But in the eyes of many Iraqis, Saddam Hussein is a hero and such clandestine activity is necessary to defend the interests of the state.

Some claim that secrecy that risks damaging the few may be justifiable in order to avoid showing cracks in the system. The Chinook military helicopter crash on the Mull of Kintyre arguably provides a trenchant example. If, as has been claimed, the accident occurred not because of human error on the part of the pilots but because of a design fault in the computer guidance system of the aircraft, far-reaching considerations come into play. The question arises: Is it better to lay blame on the individuals piloting the helicopter or to reveal that the control system on a state-of-the-art, front-line military machine may be defective? To concede the latter would open up a defensive weak spot that may give an enemy advantage, and in doing so put at risk the lives of far more personnel than two ill-fated pilots. The same argument applies to the lack of preparation for biological warfare: if the U.S. and Britain admitted having insufficient stocks of vaccines for such deadly viruses as anthrax and smallpox, a massive gaping hole in national security is revealed to potential enemies.

Likewise, the Manhattan atomic bomb project, begun in the United States in the latter part of World War II, appears to have put a sizeable number of soldiers, scientists, technicians and factory workers at risk of receiving dangerous doses of radiation without their knowledge or consent. Yet the moral argument was that such sacrifice was necessary—it permitted the development of a weapon not only designed to truncate a conflict in the Far East that threatened millions of lives, but also allowed the allied forces to stay ahead of the potentially catastrophic military advancement of Germany. In retrospect, the developments of the atom bomb helped prevent massive later conflict with the then Soviet Union.

The phrase "in the interests of national security" provides a catchall excuse for just about anything involving secrecy, whether ethically justifiable or not. Looking back with the luxury of hindsight, how can anyone vindicate the United States' incessant, long-term use of Agent Orange in Vietnam? The blanket excuse of national security has covered activities by governments and military authorities around the world that are little short of horrifying. But things could get far worse: war today is conducted in hard-hitting, short-term military offensives, using sophisticated weaponry beyond the imagination of any soldier in World War II or Vietnam. The threat of nuclear, chemical and biological war is ever present and ever increasing.

The headquarters of MI5, the UK's secret service, in London.

3.1 The Mull of Kintyre Disaster

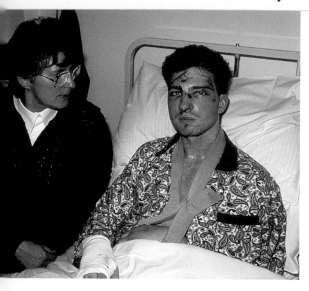

↑

Survivor of another Chinook helicopter crash (into the sea off the Shetland Isles, Scotland), Eric Morrans, recovers in hospital.

On June 2, 1994, 29 people, including some of Britain's most experienced security officers serving in Northern Ireland, plus two highly skilled Royal Air Force (RAF) pilots who had been cleared for operations with special forces, died when their Chinook Mark II helicopter plunged into a Scottish hillside in thick mist. There were no survivors. The aircraft was en route from Northern Ireland to Inverness when its flight ended tragically on the Mull of Kintyre peninsula. The official report of the RAF board of inquiry into the accident concluded that the most likely cause of the crash, based on the evidence available at the time, was that the helicopter's rate of climb was wrong, but that there was insufficient evidence to blame the pilots. Yet in spite of this conclusion, senior RAF officers, Air Chief Marshal Sir William Wratten and Air Marshal Sir John Day, chose to overrule the verdict and to attribute the cause to gross negligence on the part of the pilots, Flight Lieutenants Jonathan Tapper and Richard Cook.

Withholding evidence

The families of the pilots were outraged, but as time passed, rumors of a Ministry of Defence (MOD) coverup began to surface. By July 2000, following an investigation by journalists working for a computer magazine, evidence had come to light that the MOD had not handed over certain critical information to the RAF inquiry or to the Scottish fatal accident inquiry. Furthermore, Defence Minister Lewis Moonie had misled Members of Parliament (M.P.s), albeit unintentionally, over a nonfatal incident involving the same type of helicopter that had occurred

June 4, 1994: Rescue workers sift through the wreckage of the military Chinook helicopter, which crashed on the Mull of Kintyre on the 2nd of June.

↓

in 1989. The pilots of this aircraft, which nearly crashed, had encountered severe problems with the onboard computer system, which relied on American-manufactured software known by the acronym FADEC, or Fully Automated Digital Electronic Control.

Those campaigning to clear the names of the Chinook pilots blamed for the crash have always questioned whether it was caused by a failure in the FADEC system. Speculation about a coverup became heightened after the magazine *Computer Weekly* published an article on the 1989 accident. *Computer Weekly* claimed to have received copies of classified documents revealing an agreement between the software contractors and Boeing, the manufacturer of the Chinook, that FADEC had contributed to the 1989 crash and needed redesigning. It also alleged that the MOD had successfully sued companies in the United States, including the manufacturer of FADEC, Textron Lycoming, by demonstrating that the system had failed. The MOD, however, appears to have been seriously inconsistent in its versions of events. In 1995, the ministry had told authorities in the U.S. that the cause of the 1989 accident was "an unprecedented and violent" overrun of the Chinook engine during testing, which was blamed on the design of the FADEC system. Yet three years later, in 1998, a Defence minister told the Commons that the U.S. legal action had actually been over negligence in testing procedures.

This was not the only example of a change of story by the MOD. In submitting evidence to the U.S. courts, lawyers acting for the ministry had asserted that the FADEC software was "truly critical in maintaining safe flight." But the MOD later informed the Commons Defence Committee that Boeing "did not consider FADEC to be flight safety critical." Furthermore, neither of the two accident inquiries into the Mull crash were told that the MOD was in the process of suing Textron Lycoming over FADEC's design shortcomings.

Further damning evidence has emerged that on the day before the Mull of Kintyre accident, the MOD's own airworthiness assessors at Boscombe Down had halted test flying of Chinooks because of worries about the efficiency of the FADEC system. It has also been revealed that on June 6, just four days after the crash, an irate memo from Sir William Wratten to a senior colleague made clear that the decision of the Boscombe Down assessors was to be overruled; he also grumbled that the attitude did "nothing to engender aircrew confidence in the aircraft."

The case reopens

By the spring of 1998, independent M.P. Martin Bell called for the crash inquiry to be reopened with new evidence taken into account. This included testimony from an aviation computer expert, Malcolm Perks, and a former RAF test pilot, Squadron Leader Robert Burke. The most likely explanation for the Mull accident lay, they suggested, in two scenarios. Either a control "jam" had occurred because part of the flying controls had become detached, or the FADEC system had failed causing

 Malcolm Rifkind, former British Defence Secretary, challenged the verdict of gross negligence against the pilots of the Mull of Kintyre crash.

engine "runaway," making the aircraft virtually uncontrollable. Perks went so far as to categorize the FADEC system as "high risk." Burke revealed that while in the RAF he had been ordered not to discuss the accident. Bell pointed out that the leaked documents contradicted a claim by Prime Minister Tony Blair that the government is being "entirely straightforward and open" over the accident.

In 2000, the Royal Aeronautical Society decided that the negligence case leveled against the Chinook pilots could not be sustained when measured against the weight of evidence pointing to other technical problems with the aircraft. Following this indictment of their earlier judgment, Sir William Wratten and Sir John Day tendered their resignations from the society. It also emerged that, at the time of the original inquiry, the MOD had withheld potentially crucial information about FADEC problems from these senior RAF officers. The Ministry of Defence has since conceded, somewhat lamely, that Wratten "did not recall" the 1989 incident.

In May 2000, the MOD refused to reopen the inquiry, but in June of the same year Richard Norton-Taylor, writing for the *Guardian* newspaper, argued that the mountain of evidence that had emerged since the 1994 Mull crash had made "a mockery" of the verdict of gross negligence on the pilots. Norton-Taylor concluded that the documents obtained by *Computer Weekly* and subsequently examined by the *Guardian* exposed the reality that the MOD had repeatedly misled both the Scottish and RAF inquiries and the House of Commons. At the same time, Sir Malcolm Rifkind, the former Defence Secretary who had delivered the verdict, took the unusual step of openly challenging the earlier decision, saying that the accumulated evidence pointed to "massive uncertainty" and that it was unwise of the MOD to insist on sustaining the verdict of gross negligence against the pilots.

Fifteen coffins carrying some of those killed in the Chinook helicopter crash arrive at RAF Aldergrove on June 7, 1994.

In May 2001, in an unprecedented move, the House of Lords voted in favor of an all-party inquiry into the Mull crash. In February 2002, the committee concluded unanimously that the two air marshals should not have blamed the pilots. Growing evidence suggested that the air marshals' verdict was a clear miscarriage of justice.

The unanswered question is why did the MOD take such a position? Some people might draw the conclusion that the ministry was concerned with keeping from public scrutiny the fact that the airworthiness of a front-line military aircraft was in serious doubt and that it was better to tarnish the name of two highly qualified pilots as scapegoats.

Chinook helicopters, similar to the one that crashed on the Mull of Kintyre, use an onboard computer system known as FADEC. The reliability of this system has been severely criticised.

REFERENCES

Collins, T. Bell slams MOD over Chinook "spin" tactics. *Computer Weekly*. April 26, 2001.
www.findarticles.com/m0COW/2001_April_26/76405522/pl/article.jhtml

Fourth Report: Select Committee on Defence. *FADEC*. May 13, 1998.
www.parliament.the-stationery-office.co.uk/pa/cm199798/cmselect/cmdefence/611/df0402.htm

Minister accused over Chinook crash. *BBC News Online*. July 6, 2000.
www.news.bbc.co.uk/hi/english/uk/scotland/newsid_821000/821274.stm

New evidence on Chinook crash. *BBC News Online*. May 19, 1998.
http://news6.thdo.bbc.co.uk/low/english/uk/newsid_96000/96594.stm

Norton-Taylor, R. A fault in the system. *The Guardian*. June 20, 2000.
www.guardian.co.uk/Print/0%2C3858%2C4031465%2C00.html

Norton-Taylor, R. New doubts on cause of Mull of Kintyre helicopter crash. *The Guardian*. November 15, 2000.
www.guardian.co.uk/Print/0,3858,4091198,00.html

Norton-Taylor, R & Millar, S. Lords scorn RAF accident findings. *The Guardian*. February 6th 2002.
http://politics.guardian.co.uk/lords/story/0%2C9061%2C645658%2C00.html

3.2 The Manhattan Project

The nuclear test laboratories at Los Alamos, New Mexico, were selected for their comparative isolation from prying eyes and because their location is at least 200 miles from the nearest coastline or international border.

Victor Weisskopf, one of the most influential nuclear physicists of the twentieth century, died at the age of 93 in 2001. Of Jewish Viennese extraction and known to colleagues as the "Los Alamos Oracle," he became pivotal in one of the most consequential and secretive scientific programs of modern times, the Manhattan Project. In 1943, Weisskopf took United States citizenship. In the same year, he joined the team working on the development of the atomic bomb, which later ended the war in the Far East with terrible consequences for Japan. This experience, and the sense of responsibility that it brought, was to influence his postwar life when he resumed theoretical work on nuclear physics. Weisskopf stood diametrically opposed to the development of the hydrogen bomb, and in the postwar years he became active in the nuclear disarmament movement. From 1967 to 1973, he sat as chair of the high-energy physics advisory panel at the Atomic Energy Commission, the civilian-led successor to the Manhattan Project.

The project

In 1939, Weisskopf's experimental work on atomic physics had given him insight into the terrible potential of nuclear fission, and he was among the first scientists to argue for strict secrecy out of concern that the Axis powers might gain an advantage over the Allies in building the first atom bomb. He was not alone in taking this view. German scientists had discovered the principle and effect of nuclear fission in 1934. In August 1939, Albert Einstein, who had emigrated from Germany shortly before Hitler came to power, wrote to U.S. President Franklin Roosevelt out of concern for what might happen if the Axis powers obtained nuclear technology before the U.S. He pointed out to Roosevelt that Germany had already terminated sale of uranium from mines in Czechoslovakia and that work on uranium first begun in the U.S. was being duplicated in Berlin. In March 1942, the head of a committee charged with investigating the feasibility of an atomic bomb specifically warned Roosevelt that the United States would become involved in a nuclear arms race to develop the new weapon. It was primarily with this in mind that the U.S. president sanctioned the shadowy Manhattan Project and authorized that it be placed under utmost secrecy.

So covert was the program that even its name was the subject of careful consideration so as to minimize the chances of inadvertent exposure. It was first known under the title of "Development of Substitute Materials," but this was considered too informative and, under the leadership of General Leslie Groves, the program was then given the code name Manhattan Project, after the Manhattan District

Marines jump out of fox holes to watch an atomic cloud surge upward during one of the exercises at the atomic energy commission's Nevada proving ground.

The nuclear bomb dropped on Hiroshima almost flattened the entire city in a matter of seconds and killed thousands of people.

This undated photo of a survivor of one of the atomic blasts over Japan in 1945 shows how the checked pattern of her kimono was burned into her skin by the blast.

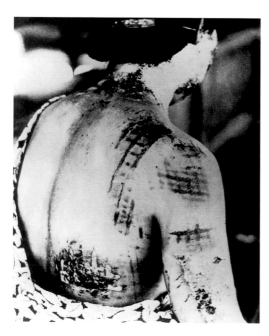

group of the U.S. Army Corps of Engineers, based in New York City, where much of the preliminary research was carried out. The bombs that the project developed also took on innocent-sounding identities including "Gadget," "Little Boy" (exploded over Hiroshima) and "Fat Man" (detonated at Nagasaki).

The people

The project team included military personnel and nearly 6,000 scientists, many of whom were German Jews who had escaped the growing mood of anti-Semitism in Europe that had begun in April 1933 with "non-Aryan" German scientists being stripped of their posts. From 1943 onward, the Manhattan team was headed by Robert Oppenheimer, a Harvard chemist who had received a doctorate in theoretical physics from Goettingen University in Germany before moving back to the United States. His main collaborators, other than Groves, Einstein and Weisskopf, included Enrico Fermi, Niels Bohr, Glen Seaborg and Leo Szilard. All were bound by a stringent code of secrecy.

The majority of later work on the program was carried out far from New York, at the site of a former boys school in the desert near Los Alamos, New Mexico. The place was chosen for several reasons: it was at least 200 miles (322 kilometers) from both the nearest coastline and the closest international border; and the area was sparsely populated. Security was unparalleled and both the scientists and their families lived in complete anonymity. Relatives could not be told of their whereabouts; all outgoing mail was screened to ensure that no detail was inadvertently revealed; and no photographs could be taken that might identify New Mexico as a location. Even the names on the team members' driving licenses were changed to numbers.

Expansion

During the subsequent Cold War period, the Manhattan Project gave rise to a crop of nuclear research facilities, all of which remained secret because the focus by then was the arms race with the Soviet Union. Laboratories were placed under strict security guard, information was classified and the patriotism of laboratory staff was under constant scrutiny. Scientists working on the nuclear weapons program operated in "classified communities" that occasionally came together in conferences to swap information under strict control. But for most of the time, a policy of "compartmentalization" kept these physicists and chemists from knowing what was going on at other research sites and even in other programs within the same laboratory. Loyalty investigations became the norm, and in its first seven years of existence the Atomic Energy Commission, created in 1946, is said to have investigated half a million people for security clearance at a cost of more than $10 million a year.

Yet the Manhattan Project, and the intense secrecy it engendered, acted as a springboard for change. As early as 1945, Weisskopf became a founding member of the Federation of Atomic Scientists, an organization not only committed to warning of the consequences of nuclear war but also specifically working to reduce the scope of government secrecy through pressure to accelerate the declassification of Cold War documents.

The dropping of nuclear bombs in Japan led to their surrender in World War II. Here, Namoro Shigomitso, aboard the battleship *Missouri*, signs the Japanese surrender document on behalf of his government and the Emperor.

REFERENCES

Aftergood, S. Project on government secrecy. *Federation of American Scientists.* Undated.
www.fas.org/sgp/aftergood.html

Dowling, M. The Manhattan Project. Last updated May 1, 2002.
www.mrdowling.com/706-manhattanproject.html

The Manhattan Project-Appendix: Key figures in the Manhattan Project. *Undergraduate Engineering Review, University of Texas.*
www.me.utexas.edu/~uer/manhattan/people.html

The Manhattan Project-introduction. *Undergraduate Engineering Review, University of Texas.*
www.me.utexas.edu/~uer/manhattan/intro.html

Westwick, P. J. 2000, November/December. In the beginning: the origin of nuclear secrecy. *Bulletin of the Atomic Scientists 56(6).*
www.bullatomsci.org/issies/2000/nd00/nd00westwick.html

3.3 Smallpox: How Ready Are We?

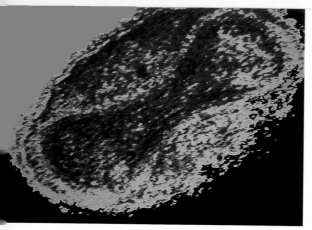

The deadly smallpox virus, magnified 120,000 times.

In November 2001, the Bush administration in the U.S. reached a decision to enlarge its stockpile of smallpox vaccine so that, in the event of a terrorist attack or any other use of smallpox as a biological weapon, the Centers for Disease Control and Prevention would be able, in theory at least, to vaccinate everyone in the United States. Intelligence reports suggest that some samples of smallpox from the Russian State Research Center of Virology and Biotechnology, one of two remaining repositories in the world, may have been illegally supplied to terrorists. Russia, it is said, still possesses an industrial facility that is capable of producing tons of smallpox virus annually and maintains a research program thought to be working on more virulent and contagious strains.

In the U.S.

U.S. Health and Human Services secretary Tommy Thompson noted that the risk is low, but that it exists and America "must be prepared." Smallpox vaccination has not been routinely carried out in the U.S. in the last 25 years and immunity acquired before that time has undoubtedly lost much of its effectiveness. CNN has reported that the administration currently possesses enough vaccine in the short-term to produce some 77 million doses by diluting existing stockpiles, though it is not made clear what effect dilution might have on the efficacy of the treatment. In 2000, a contract was awarded to a Cambridge, Massachusetts, firm, Oravax, to produce 40 million doses of vaccine with anticipated delivery of the first full-scale consignments in 2004. The latest agreement, worth $428 million, gives the go-ahead to a British company, Acambis, for the production of a further 155 million doses, with the option to buy up to 500 million doses.

Bob Stevens, a British man who lived in Florida and worked for *The Sun* newspaper, died of anthrax on October 5, 2001 after antibiotics failed. He was the first person to die of anthrax in the U.S. for 25 years.

Across the water

The news of U.S. activity on smallpox containment may be set beside a February 2002 communication from Pat Troop, the deputy chief medical officer for the British Department of Health. This was put out as an urgent Public Health Link prior to a BBC2 Television drama-documentary focusing on a fictional large-scale bioterrorist attack using the smallpox virus. Troop offered the vague assurance that the Department of Health has "a substantial stock of smallpox vaccine, manufactured in the 1970s...information about the size and location of vaccine stocks is not being put in the public domain as this is information that might be of use to terrorists. These stocks could be rapidly deployed to contain an outbreak."

Could they? What size of outbreak does the U.K. administration anticipate containing? These are fair questions because nowhere in her

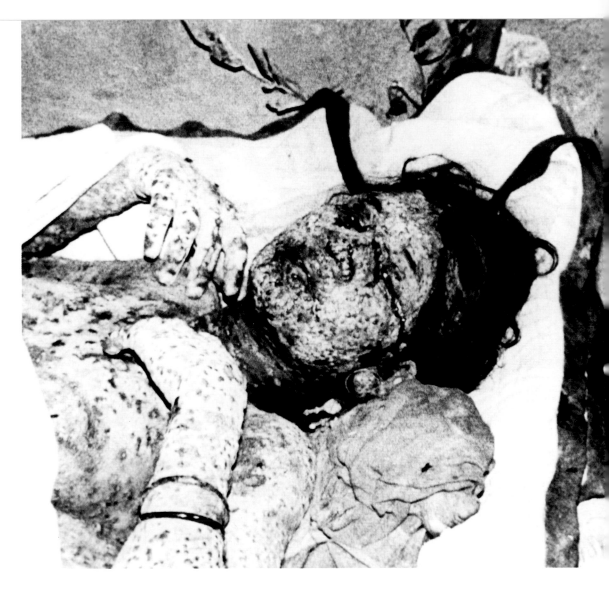

A disfigured woman is seen in hospital in 1974 dying of smallpox in the state of Bihar, India. Emergency vaccinations were conducted in the area, where an estimated 25,000 had already died of the disease.

communiqué does Troop disclose more precise facts and figures. In 1999, an editorial in the British medical journal *Lancet* reminded how fast smallpox can spread once liberated into the environment. Vaccination is no longer routine, since the last outbreak occurred in Somalia in 1977 and smallpox was declared eradicated by the World Health Organization in 1980. Currently, the U.K. health policy is only to give smallpox vaccinations to laboratory workers who might come into contact with the virus.

In 1999, a leading medical journalist warned BBC Television that stocks of smallpox vaccine were too low, and a spokesman for the Public Health Laboratory Service (PHLS), the authority that in theory would distribute the vaccine, said "The PHLS keeps about 70 to 80 doses in stock. In the case of a bioterrorist attack it would be up to the Department of Health and Home Office to formulate a plan." Were a

October 25, 2001: Hazardous
materials experts work in the
streets surrounding Capitol Hill,
Washington, D.C., looking for
traces of anthrax contamination.

fresh outbreak to happen, given the present stock levels, it would take up to three years to manufacture enough vaccine for an effective vaccination programme. In 2002, it was revealed that the British government was to buy up a £32 million ($51 million) stockpile of smallpox vaccine. Yet, according to a *Daily Telegraph* report, Pat Troop refused to tell a Commons Defence committee hearing how many doses of the vaccine had been ordered, on the grounds that the information might be useful to an enemy. An irate Labour M.P. on the committee, Kevan Jones, responded: "You are insulting my intelligence. I don't accept that is a national security issue."

REFERENCES

Center for Civilian Biodefense Strategies, John Hopkins University.
Smallpox. Copyright 2002
**http://www.hopkins-biodefense.org/pages/agents/
agentssmallpox.html**

US Department of Health and Human Services. HHS awards $428
million contract to produce smallpox vaccine. Revised November
28th 2001.
http://www.hhs.gov/news/press/2001pres/20011128.html

BBC News. Smallpox vaccine warning. May 7th 1999.
**http://www.bbc.co.uk/hi/english/health/newsid_337000/
337289.stm**

UK Department of Health. Urgent communications re. BBC
Programme on a fictional smallpox outbreak, Tuesday 5 February
9.00pm—Smallpox 2002—Silent Weapon. February 4th 2002.
http://www.doh.gov.uk/cmo/cmo02_01.htm

3.4 Agent Orange

One of the most shadowy issues surrounding the Vietnam War era and its aftermath concerns a defoliant with the innocuous-sounding code name Agent Orange. Why this title? In order to disguise the true nature of the substance, it was identified only by the orange color bands painted on the 55-gallon drums in which it was supplied to the U.S. military. Agent Orange represents only one of 15 herbicides used in Vietnam from the early 1960s (all color coded) until its use was discontinued in 1971, but its involvement was by far and away the most consequential.

The chemical

Agent Orange was first produced and tested in the 1940s at various military establishments in the United States. These included Fort Detrick in Maryland, Eglin Air Force Base in Florida and Camp Drum in New York. Later, the defoliant was tested in Thailand. Its military application came into effect only during the conflict in Vietnam, where it is estimated some 19 million gallons were eventually dispersed either deliberately or inadvertently through accidental spillage. The agent was sprayed from aircraft, mainly C-123s, land vehicles and small containers suitable for hand application. It was manufactured for the U.S. military by a number of companies including Dow, Monsanto, Diamond Shamrock Corporation, Hercules Inc., Uniroyal Inc., T-H Agricultural and Nutrition Company and Thompson Chemicals Corporation.

1966: Two United States planes of the 12th Air Squadron fly low over Vietnamese jungle spraying Agent Orange.

These two photos show the environmental devastation that Agent Orange caused. The top photo shows a mangrove forest near Saigon before Vietnam was sprayed. The bottom photo is of a mangrove forest in 1970, five years after the area was sprayed with Agent Orange by U.S. planes. The dark spots are some surviving trees.

The consequences

The defoliant consists of a mix of two chemicals in roughly equal proportions: 2,4-Dichlorophenoxyacetic acid (known as 2,4-D) and 2,4,5-Trichlorophenoxyacetic acid (known as 2,4,5-T). For application, this combination was then added to kerosene or diesel fuel. It was found to be extremely effective in killing off the foliage of broadleaf trees and shrubs in the dense jungle terrain typical of Southeast Asia, where the enemy could hide. Outside of the manufacturing companies, it was not generally known, however, that Agent Orange was also highly contaminated with a substance called TCDD. This unwanted by-product of the chemical manufacturing process is one of a group of intensely poisonous compounds, dioxins, which are proven to cause an assortment of illnesses in humans, many of them fatal.

Military personnel serving in Vietnam were not, it is alleged, generally aware of the dangers of working in jungle environments in which Agent Orange was being used, and the United States Code prevents veterans from suing authorities for injuries that may have resulted. Dioxin analysis of blood samples, a fairly expensive process (said to cost up to $2,000 per test), has never been made generally available on a large-scale basis, and the U.S. Department of Veterans Affairs does not perform such clinical examinations. Nevertheless, a lawsuit was filed in 1979 against the manufacturing companies, under the designation of product liability litigation, and an out-of-court settlement of $180 million was reached in 1987. This, however, appears to fall a long way short of reflecting the full story.

The effects continue

In the summer of 2000, it was reported that villagers living in the proximity of Bien Hoa, a former U.S. airbase in Vietnam where there had been a substantial accidental spillage of Agent Orange, were still showing raised levels of dioxin contamination. It is thought that the spillage escaped into a nearby watercourse and contaminated the silt, which in turn contaminated the fish that constitute a major proportion of the local people's diet.

According to Arnold Schechter, a professor of environmental sciences at the University of Texas and former member of the U.S. Army Medical Corps in the Vietnam War, dioxin remains in the food chain in various "hotspots" in Vietnam. Schechter reported that a woman from Hanoi

from whom a blood sample was taken for testing possessed a dioxin level that represented a 135 percent increase over the level recorded for Hanoi residents who had not been exposed to the agent. Yet in 2000, almost 30 years after the spraying of jungle in North Vietnam had ended, no U.S. government funding had been set aside for research into the consequences of Agent Orange use in Southeast Asia. It has been pointed out that such research is urgently needed to establish how dioxin passes through the environment and is able to contaminate people 30 years after exposure. As Schechter says, the implications extend beyond Vietnam and U.S. veterans seeking compensation for exposure to Agent Orange.

Schechter commented recently, "Vietnam has the biggest dioxin contamination in the world and probably the most men, women and children contaminated with dioxin." His is one of the pieces of bad news

In Tu Du Hospital, southeast Asia's largest obstetrics hospital, five gallon jars are filled with stillborn or aborted fetuses. Most are deformed. According to the Vietnamese medical authorities, each fetus is a possible victim of dioxin poisoning from Agent Orange. As many as ten pairs of Siamese twins are born here each year, where as one case every ten to 15 years would be expected.

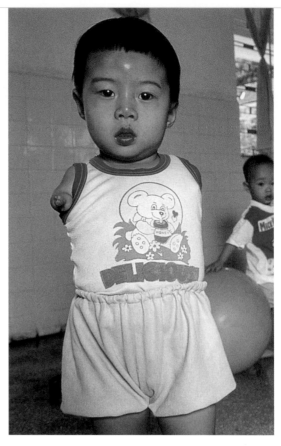

A child suffering the genetic effects of Agent Orange. His parents were never actually sprayed but instead infected by the water in fruit.

At Saigon's Tu Du hospital, children born with defects play in a ward.

from the Vietnam era that seems not to have been widely publicized by the U.S. government. Schechter's observations need to be seen in light of the fact that by 2000, the U.S. Environmental Protection Agency had re-assessed its evaluation of dioxin and concluded that it is a carcinogen that causes cancer in humans.

Meanwhile, since 1992, when the Supreme Court refused to hear arguments from the plaintiffs on the principle that "the matter is settled," further attempts to sue manufacturers have been prohibited by U.S. courts.

REFERENCES

Frequently asked questions about Agent Orange. *The Agent Orange Website*. Undated.
www.kewispublishing.com/faq.htm

Overview. *The Agent Orange Website*. Undated.
www.kewispublishing.com/orange.htm

Reuters. Agent Orange still hot in Vietnam. *Wired News*.
July 12, 2000.
www.wired.com/news/print/0,1294,37521,00.html

3.5 Radiation Exposure

John Wayne and Susan Hayward filming *The Conqueror* 100 miles from the Nevada nuclear test site. Both Wayne and Hayward, as well as half of the crew, contracted cancers in subsequent years.

In 1954, a film starring John Wayne, entitled The Conqueror, was made on location some 100 miles (160 kilometers) downwind of the United States nuclear test site in Nevada. Almost half of the people involved in production of the film, including Wayne, subsequently contracted cancers. A link between radioactive contamination and the fate of the film workers was sought but never proved. A little more than 20 years later, in November 1974, Karen Silkwood, a chemical technician at the Kerr-McGee plutonium fuels production plant in Crescent, Oklahoma, died in an unexplained car crash involving no other vehicle. Reportedly, in the week prior to the accident she was gathering evidence in support of allegations that the company had been negligent over plant safety and the exposure of workers to plutonium radiation. Out-of-court compensation of $1.3 million was eventually settled 12 years after her death, but the circumstances remain a matter of speculation. These two well-publicized cases represent the tip of an iceberg, the existence of which was deliberately "buried" for decades by the U.S. government in the form of classified reports.

February 2002, however, saw a damning exposure of U.S. official secrecy and disregard for human safety during the period of Cold War experimentation with nuclear weapons and radioactive substances. The Web site of the prestigious U.S. Institute for Energy and Environmental Research (IEER) published a progress report on a long-term feasibility study conducted by the National Cancer Institute and the Department of Health and Human Services Centers for Disease Control and Prevention. The study, commissioned in 1998 by U.S. Congress in the wake of mounting public pressure for the disclosure of facts, examines the health consequences of fallout from nuclear tests in the United States, Britain, France and the Soviet Union.

On May 25, 1953, military top brass and U.S. Congressmen unwittingly risked radiation exposure as they observed the detonation of the first atomic artillery shell.

Official publication of the study has thus far been delayed, but extracts obtained by Iowa Democratic senator Tom Harkin have been released through IEER. Harkin cites passages from the report revealing that any person alive in the United States since 1951 has been exposed to radioactive fallout and that all organs and tissues of the body have received radiation exposure. The feasibility study estimates that about 22,000 cancer cases among U.S. residents alive between 1951 and 2000 will occur as a result of fallout exposure and that about 11,000 extra deaths will result. The estimate is based on theoretical radiation doses from the Nevada test site and from testing outside the United States. Analysis by the IEER of the government studies, however, suggests that about 80,000 people living in the United States during that period will contract or have contracted cancer as a result of nuclear test fallout and that well over 15,000 of these cases will prove fatal.

Biologist and IEER outreach director for the United States, Lisa Ledwidge, has commented, "We applaud the fact that the United States government has been honest enough to say that it has harmed its own people, though it did so only under prolonged pressure from the people and some of its elected representatives...the harm is still occurring. The government needs to inform people fully." One of the examples of official secrecy revealed by the IEER indicates that in the 1950s, the U.S. government advised photographic film manufacturers to protect their stocks from predicted fallout, but did nothing to inform American milk producers so that they could protect an essential food supply.

Hazardous working conditions

In May 2000, *USA Today* published details of a secret program that turns attention back to the type of evidence Karen Silkwood was gathering before her death. In the 1940s and 1950s, the U.S. government recruited large number of private companies, mostly in the industrial belt of New England, New York, New Jersey, Pennsylvania and Ohio, to process nuclear weapons material. Recently declassified documents reveal that during the development of the atomic bomb, the U.S. government hired more than 200 companies to quietly switch to the processing of nuclear weapons material. At least one-third of them handled radioactive and toxic substances, often without the equipment or knowledge to protect either workers or local communities. The contracts were largely kept secret, as were the dangers when thousands

of workers were exposed to levels of radiation hundreds of times above accepted safety standards.

Federal officials, it is alleged, knew of severe health risks to industrial workers and technicians engaged in building the U.S.'s early nuclear stockpile, but misled them by telling them, contrary to clear evidence, that they were operating in safe conditions. It is said that vital information was never published because of fears that employees in the plants would seek "danger money," sue for damages or demand safer working conditions.

In one particular case cited by *USA Today*, a former medical director with the U.S. Atomic Energy Commission, Bernard Wolf, had incontrovertible evidence of serious dangers to which workers processing uranium at the Harshaw Chemical Company in Cleveland, Ohio, were exposed. Classified tests by his staff had revealed airborne concentrations of uranium dust 200 times higher than the safety limits accepted today. But when Wolf visited the plant in January 1948, he assured the employees that the uranium posed no threat to their health. Reports examining the problems in many of the factories producing weapons-grade material were classified and locked away, and in some instances the secrecy is still maintained. As Dan Guttman, a former director of the President's Advisory Committee on Human Radiation Experiments, set up in 1994, has observed, "These places just fell off the map."

Standing their ground

It has been reported that surviving employees still have not been properly informed of the risks they carry, even though screening could increase their chances of survival. Dan Guttman has stated that the government knew safety standards were being abused, but there had been no effort to maintain contact with the people put at risk and to look at the effects.

Employee of the Kerr-McGee Nuclear Corporation, Karen Silkwood, died in a fatal but unexplained car crash near Oklahoma City on November 13, 1974. In May 1979, a U.S. federal jury found the corporation liable for her radiation contamination and awarded her estate more than $10 million. But was her death an accident?

Workers in today's nuclear power plants wear head-to-toe protective clothing to prevent radiation from entering their bodies. In the early days, the knowledge about the effects of radiation was much lower, as was the standard of safety.

Communities have been left in ignorance of radioactive waste on their doorsteps, and secrecy still obscures the facts about contamination at some sites. In the early 1970s, the U.S. Congress created a body known as FUSRAP (Formerly Utilized Sites Remedial Action Program) to identify and clean up contamination at sites where commercial organizations carried out nuclear weapons work. But after 25 years, only 28 out of the 46 sites earmarked for action have been cleaned up. Furthermore, some of those for which a cleanup was originally considered unnecessary are now on the list. These include the Harshaw site, deemed ineligible for the FUSRAP program in 1978. Since then it has been derelict, fenced off and smothered in radiation hazard notices.

The situation is summed up by an 87-year-old former Harshaw employee, John Smith, who expressed the sentiment to *USA Today* reporter Peter Eisler, "It makes you bitter, them knowing about the risks and not telling. If I'd known I would have quit."

REFERENCES

Eisler, P. Secret program left toxic legacy. *USA Today.* September 5, 2000.
www.usatoday.com/news/poison/001.htm

Eisler, P. Worker risks weren't a priority. *USA Today.* September 6, 2000.
www.usatoday.com/news/poison/008.htm

Harkin, T. Progress report: a feasibility study of the health consequences to the American population of nuclear weapons tests conducted by the United States and other nations. *Official Documents. August 2001. IEER.* February 6, 2002.
http://www.ieer.org/offdocs/index.html

IEER press release re the government study referenced above. *IEER.* February 28, 2002.
www.ieer.org/comments/fallpout/pr0202.html

Makhijani, A. A global truth commission on health and environmental damage from nuclear weapons production. *IEER.* February 2001.
www.ieer.org/sdafiles/vol_9/9-2/truth.html

Poisoned workers and poisoned places. *USA Today.* June 24, 2001.
www.usatoday.com/news/poison/cover.htm

Simpson, J. and S. O'Brian. Still waiting. *USA Today.* September 7, 2000.
www.usatoday.com/news/poison/018.htm

3.6 UFOs: A Government Fabrication?

Unidentified Flying Objects (UFOs) have captured the imagination of the public in the past century, but the real truth may be far different from what is commonly believed. In recent years, under mounting public pressure, the U.S. Central Intelligence Agency (CIA) has declassified a limited number of files relating to its historical interest in UFOs, while in 2000 the U.K. Ministry of Defence (MOD) handed over hitherto classified information to Colin Ridyard, a chemist working for public interest group.

It is far from clear, however, exactly how much information has been withheld since the UFO phenomenon began in the late 1940s. In 1966, public pressure in the U.S. resulted in a study being commissioned into flying saucers under Dr. Edward Condon, a University of Colorado physicist. The report, issued in 1969, concluded that little, if anything, had come from the study of UFOs and that further extensive study was unwarranted. Interest groups regarded the report as no less than a coverup and belief grew that the CIA was involved in a major conspiracy.

Similar concern about MOD secrecy has prevailed in the U.K. As Colin Ridyard commented recently in an *Observer* newspaper interview, "It is clear that there are many strange incidents that happen in the British skies that are kept secret. There may be issues of aircraft safety or natural phenomena, but by keeping this information secret these incidents cannot be scrutinized by the public or the scientific community."

A double bluff

Cold War paranoia about spy planes and national security seems to have influenced Western governments. In 1996, it was revealed in an article for *Studies in Intelligence*, a formerly classified CIA publication, that false explanations were regularly concocted by national security officials to disguise the fact that "UFO detection" frequently amounted to the sighting of secret U2 spy planes. The author of the report, Gerald Haines, claims that more than half of the UFO reports during the 1950s and 1960s were sightings of spy planes. It has been claimed that many other UFO and "alien" sightings were actually nothing more than Air Force balloon flights and impact tests, wherein dummies were dropped into the desert. Haines, however, advised U.S. publication *Dossier* that the CIA had removed a limited amount of information from his report before it was released, although these classified sections "had nothing to do with UFOs" directly.

It was the famous 1947 Roswell incident in New Mexico that stimulated real interest in UFOs, but there was almost certainly another factor. At that time, the lives of ordinary people were overshadowed by the fear of nuclear holocaust between East and West. The coauthor of a recent study into the UFO phenomenon concluded that people wanted to believe in something "up there in the sky that could come and rescue us."

June 1997: The U.S. Air Force released this 1972 photo of a Viking space probe awaiting recovery at White Sands Missile Range in New Mexico as part of its report on the so called Roswell Incident of 1947. The Air Force reported that 'space aliens' who suposedly crashed in the New Mexico desert 50 years ago were only military dummies and that high-altitude research projects such as this may have become part of the incident. The report is aimed at ending longstanding speculation over the incident and denies that the military had recovered bodies from damaged flying saucers in 1947.

Dr. Edward Condon, the University of Colorado physicist who led the 1960s research into flying saucers.

In January 1953, the CIA commissioned a secret report by a group of experts, known as the Robertson Panel, intended to assess the national security implications of UFOs. Gerald Haines reveals that the CIA made desperate efforts to keep the Robertson documents classified. An editorial in the *Washington Post* of August 8, 1997, described the difficulties of getting internal documents on UFOs to be "like pulling teeth." Commenting on the implications of the Haines report, intelligence author David Wise asserted, "You don't have to believe in little green men to see the admitted deception as yet another example of official lying that has eroded public trust in government." When challenged about official deceit over UFO sightings, U.S. Air Force spokesman Brigadier General Ronald Sconyers responded, "I cannot confirm or deny that we lied."

During the height of the Cold War, research was allegedly carried out by both the Americans and the Soviets on the development of flying saucers. It has been suggested that the CIA and the MOD concealed this out of concern that it might trigger national hysteria over the Soviet Union's possible "super weapons." The CIA's argument has been that the intelligence programs on "unusual observations" were kept secret "to protect national security assets." A similar story prevailed in Britain where, according to declassified documents from 1965, "it was official MOD policy to play down the subject of unidentified flying objects." At least one explanation for this reticence may be that British and U.S. defense authorities were anxious to hide deficiencies in their radar systems during the early postwar years, when radar equipment was still in its infancy and may have been unable to recognize state-of-the-art Soviet military aircraft.

The truth is out there

In 1999, the British Civil Aviation Authority reported two incidents to the MOD, including the radar detection of a large object traveling over the Scottish coast at 3,000 miles per hour. In the other, the pilots of four aircraft reported seeing a "ball of light" moving at high speed over the North Sea. Air traffic controllers reported no strange aircraft, but a weather station radar operator tracked a fast-moving object in the area. An MOD spokeswoman claimed to the *Observer* that incidents of that kind all had normal explanations. But a U.S. Air Force spokespersons has candidly stated, "We have classified programs that we need to protect."

REFERENCES

Barnett, A. UFO sighting claims. *The Observer.* June 4, 2000.

Elliston, J. The UFO Cold War: New revelations in a CIA-published study. *ParaScope.* 1997.
www.parascope.com/articles/0897/ufolies.htm

Haines, G. K. CIA's role in the study of UFOs, 1947-90. *Central Intelligence Agency.* 1997.
www.cia.gov/csi/studies/97unclas/ufo.html

Harris, P. Cold War hysteria sparked UFO obsession, study finds. *The Observer.* May 8, 2002.

U.S. lied to explain away UFO sightings, says CIA. *Institute for UFO Research.* 1997.
www.frii.com/~iufor/cia_ufo.htm.

Rodeghier, M. The CIA's UFO History. *CUFOS.* Undated.
http://www.cufos.org/IUR_article3.html

4 Power Corrupts

Political gain takes precedence over public interests

4 Power Corrupts

Politics and secrecy often go hand-in-glove. Newspapers perennially reveal to us that governments and their agencies have been doing things behind our backs, often activities we would not willingly condone. In December 1998, the U.K. director of the Campaign for Freedom of Information commented on the parliamentary ombudsman's report on access to official information by saying, "You could not ask for a clearer demonstration of the indelibly ingrained secrecy of the British authorities than this report." Much the same can be said for the attitudes of governments in many countries.

↑

The level of government secrecy in the U.S. is said to have deepened since the inauguration of President George W. Bush.

In a democratic, political party-led country, politicians need to convince the electorate that their policies are in the interests of the citizens of that nation, that they are representing those people's views and acting accordingly, regardless of whether they actually are or not. An admittance of error, incompetence or bad judgment in any area of politics or economics is an admittance of ineffectiveness at the job and will likely be heeded by the voters at the next elections.

Some issues are so ingrained in our society, our industry and our economic success that they seem, on the surface, irreversible. In the case of genetically modified (GM) foods, governments in the West sold the idea to people that we could feed the world by developing pest-resistant crops that are bigger and more bountiful than their natural predecessors. The intention may have been good, but the reality may not be. The results of research by both anti- and pro-GM lobbies have been manipulated and withheld, creating confusion amongst consumers. Do politicians know something we don't? If they did, would they tell us? A withdrawal of all GM crops around the world after so many years of production would have devastating effects on the world's resources and the world economy. What better reason is there for governments to keep quiet?

Spin doctors have become part of the political furniture in the U.S. and Britain; their job is to turn any issue around to shed rosy light on the politicians they represent. It is no wonder, then, that secrecy, coverups and misleading information have become part-and-parcel of modern politics.

Nowhere is this more true than in issues with a scientific basis. Politicians make judgments on the advice of medical institutions, scientific researchers and large companies. Legislators are rarely specialists in any field, particularly today when so many are career politicians with no experience of industry, so their decisions are based on advice. But where does this advice come from? Whose interests are the advisors serving? In more cases than not, the advisors work for multinationals, who pay huge donations to political party coffers in the hope that their interests will be taken into account when politicians make their decisions. Political parties know on which side their bread is buttered and where the funding for their next electoral fight will come from, so it is no wonder that the advice of companies is often taken.

But this is not always the reason for government secrecy. The advice may be good, it may be in the interests of the nation and its citizens, but the policy put into place may go against it. Since the first days of world empires, invasions and international trade, the economics of nation states has been at the forefront of political decision-making. State or local authority officials may ride roughshod over environmental concerns out of a desire to gain prestige, wealth or both.

Nuclear power was heralded, in its heyday, as the energy source of the future, and even today is purported by the international nuclear industry as the only means by which sustainable development will be achieved. Despite two major incidents, at Three Mile Island and Chernobyl, and many minor accidents, the number of nuclear power stations is still increasing and the world's reliance on nuclear power is deepening. International organizations, individual governments and the international nuclear industry will not back down on nuclear power in favor of alternative sources. The reasons for the denial are clear: in Britain, 25 percent of the nation's electrical power comes from the 15 nuclear power stations around the country, to which British Nuclear Fuel hopes to add 10 more. In the world as a whole, 17 percent of all electricity comes from nuclear power, an ever-increasing percentage as more and more countries (such as North Korea in 2002) add to their reliance on nuclear power. If governments fully admitted that nuclear power stations were a danger to the people who lived nearby, that their by-products were detrimental to the environment and that their waste cannot be suitably treated and disposed of, the effects on the world economy would be disastrous.

The core of a nuclear reactor, underwater, during the charging period. Minor accidents and major disasters scatter the history of nuclear power. But is the past just the tip of the iceberg? Are there more catastrophes waiting to happen? And what will happen to the thousands of tonnes of nuclear waste that needs to be isolated from the biosphere for tens of thousands of years.

REFERENCES

Ombudsman report reveals "indelibly ingrained secrecy" of public authorities. The Campaign for Freedom of Information. December 10, 1998. **www.cfoi.org.uk/ombud101298pr.html**

4.1 Nuclear Power Plant Disasters

Three Mile Island nuclear power plant, Pennsylvania, where a more extensive catastrophe was only narrowly averted by engineers working there.

Arial view of the ceiling of the remains of the core of Reactor 4 at the Chernobyl nuclear power station. The covering was built to prevent further release of radioactive contamination from the damaged reactor. In 1998, the Deputy Director of the plant, Valentin Kupny, said that the concrete sarcophagus is in extremely bad condition and could collapse at any time.

The first nuclear reactors were built to meet the demands of the arms race and to provide weapons-grade plutonium. Concealment played its part from the outset, since in order to disguise the exact purpose for which such facilities were intended, they were euphemistically labeled "production reactors." A whole series of such plants went into service in the United States, Russia and the United Kingdom after World War II, but Britain then led the way to a fundamental extension in the use of nuclear facilities, which saw the advent of the nuclear power station as we now know it. In the 1950s, Calder Hall, in the northwest of England, opened with the purpose of not only producing weapons-grade material but also using the steam generated as a by-product from the reactor cooling water to drive turbines and thus inject power into the national grid. By and large, strict safety procedures and careful design meant that these installations ran well. But the potential for an accident has always been present and disaster has struck twice. It happened first in the United States and then, with more catastrophic consequences, in the Soviet Union.

Three Mile Island, U.S.A.

In the early hours of the morning of March 28, 1979, an unexpected problem arose in the twin reactor located on Three Mile Island in the Susquehanna River near Harrisburg, Pennsylvania, a town lying about 90 miles west of Philadelphia. An automatic valve facilitating the supply of cooling water to the Number 1 reactor shut down, which meant that cooling water was prevented from reaching the reactor core. Safety

Colored satellite image of the area around the Chernobyl nuclear power facility. The red areas denote vegetation. The large black area at the centre is the power station's 7.5-mile (12-km) long cooling pond. The power station complex is at the left end of the pond with the site of the four RBMK-type nuclear reactors seen as a dark brown rectangular area. An exclusion zone of 18.5 miles (30km) in radius around the site covers most of this frame.

devices came into play that shut down the reactor to save it from overheating, but the fault was not identified properly. A series of human errors and equipment failures then compounded the original problem resulting in superheated steam, by that time circulating freely in the system, coming into direct contact with the reactor core, half of which went into meltdown. The fuel had been clad in a material called zirconium and this reacted with the steam to generate radioactive hydrogen gas, some of which escaped into the containment vessel surrounding the reactor. Fortunately, the staff at the plant reacted quickly to restore the circulation of cooling water and very little of the gas escaped into the atmosphere.

The Three Mile Island incident could have resulted in catastrophe and it served as a timely warning. In the United States, nuclear power came under fierce public protest and, in response, seven other reactors of similar design were shut down. Licenses for new reactors were put on hold and it was not until the mid-1980s that any were actually ordered by the nuclear industry in the U.S. Only in 1985 did the undamaged reactor at Three Mile Island restart; the Number 1 reactor remains shut down to this day.

Chernobyl, former Soviet Union

The warning was not, unfortunately, heeded everywhere. The potential for nuclear accident remained in place in other parts of the world, where the lessons had not been heeded and where power stations were less well designed and often maintained by poorly trained staff.

Such an accident happened in 1986 on the other side of the world. At Chernobyl, in what is now the Ukraine, engineers had built a nuclear power station, which included four reactors, by adapting an antiquated design based on the reactors that were employed in Siberia to produce weapons-grade material. The Chernobyl plant was rickety, with unacceptable levels of safety that were concealed not only from the outside world but also from the people living in its shadow. The reactors were controlled by the manipulation of graphite rods. When these were fully inserted into the reactor core, they served to slow down the process of the nuclear reaction, but if they were partially withdrawn, the process speeded up. The heat of the reaction produced steam, which drove turbines to generate electricity.

On the night of April 25, 1986, the Chernobyl engineers were conducting an experiment that was to prove fatal. They set out to establish whether the turbines would still generate electricity after the

December 1995: Abandoned secondary school in the exclusion zone around Chernobyl, Ukraine.

supply of steam was shut off and when running down from full speed. To carry out this test, reactor Number 4 was to be run at low power with half the number of control rods inside the core. Under the circumstances, the technicians felt confident about disabling the automatic shut-down system and the emergency cooling system while controlling the reactor manually. When the steam to the turbines was turned off, however, the pressure inside the reactor rose and the cooling water began to boil, resulting in a surge in power output. Although a technician hit the emergency button within 30 seconds of the needles starting to climb off the dials, it was too late for the remaining control rods to be re-inserted and the reaction rapidly became unstable. The mounting steam pressure blew off the roof of the reactor allowing in air and causing the graphite rods, which by this juncture had become white-hot, to ignite. The resulting fireball blasted a cloud of highly radioactive material more than half a mile (one kilometer) into the atmosphere. Winds then blew the contamination all over Europe.

To put out the fires required the dumping of 5,000 metric tons of rock and lead from helicopters; the whole reactor was eventually sealed in concrete. More than 750,000 workers had to be drafted for the cleanup operation, most of them unaware of the severe risks to their health. During the weeks after the accident, the Soviet authorities evacuated 135,000 people and banned all civilians from an 18.5-mile (30-kilometer)

April 8, 1998: A worker checks the radiation level at the command console of Chernobyl's Reactor 4.

Two workers wearing protective clothing and oxygen masks before they undertake decontamination work at Three Mile Island nuclear power plant. The cleanup operation started in 1989 and is expected to go on until 2029.

radius of the plant. It has been calculated, however, that nearly half a million people were exposed to unacceptable levels of radioactivity and up to 15,000, including 31 members of the original station staff, died from radiation-linked diseases in the decade following the Chernobyl explosion. One of the longer-term consequences has been a rise in the incidence of thyroid cancers.

The danger is by no means over. The concrete cladding around the damaged reactor has apparently now leaked and there is a distinct possibility that the roof will collapse. Many people are perhaps unaware that two of the four original reactors are still in service producing electricity. Furthermore, many Chernobyl-style reactors, operating under the same design shortfalls, remain functional across Eastern Europe in the countries of the former Soviet bloc. Economic problems that have arisen since the collapse of Soviet communism also mean that engineers and technicians are often inadequately paid, and maintenance and safety procedures are lax.

REFERENCES

Chernobyl—The Accident (factsheet). *Bellona Foundation*. Undated
http://www.bellona.no/imaker?id=12663&sub=1

Chernobyl Ten Years On: Radiological and Health Impact (Chapter 1: The Site and Accident Sequence). *Nuclear Energy Agency, France*. Undated.
http://www.nea.fr/html/rp/chernobyl/c01.html

The Three Mile Island Recovery and Decontamination Collection. *Penn State University Libraries*.
http://www.libraries.psu.edu/crsweb/tmi/tmi.htm

Three Mile Island Special Report. Washingtonpost.com. March 27, 1999.
http://www.washingtonpost.com/ wp-srv/national/longterm/tmi/tmi.htm

4.2 Is There a Nuclear Power Plant Coverup?

A worker at Sellafield stands in front of the charge face of Windscale Pile One, the scene of Britain's worst nuclear incident in October 1957 when a nuclear fire resulted in shutdown.

It may seem like a long time since the Three Mile Island nuclear accident and the disaster at Chernobyl, so it would be fair to assume that the lesson must surely have been learned, and that since it looks as though nuclear energy is here to stay, safety precautions are regarded as paramount. Apparently this is not always the case, and complacency, lax management and coverups may still be the order of the day in British nuclear installations and perhaps elsewhere. In April 2002, one of the members of a U.K. government panel examining the effects of radiation, Dr. Chris Busby, accused Somerset health officials of deliberately misleading the public about radiation effects from power stations in the west of England.

In a report commissioned by the Green Audit organization and published in April 2000, Busby, with coauthors Paul Dorfman and Helen Rowe, identified 61 women who suffered from breast cancer between 1988 and 1997 in Burnham, a town only four miles from the nuclear power plant at Hinkley Point on the north Somerset coast. He concluded that this figure was abnormally high. While the national statistic offered a projection of 8.7 cases of breast cancer in a population area of comparable size, 17 deaths were recorded in Burnham North, directly downwind of the power station.

The county health authority promptly described the report as "alarmist," claiming that the doctor had used "poor maths." A spokesman for the Somerset Coast Primary Care Trust insisted that there was no evidence linking Hinkley to the high incidence of breast cancer. He asserted, "To say public health doctors would have ulterior motives other than the health and well-being of the community is utterly preposterous." Busby's criticism, however, brought to light the fact that officials had taken his figures and percentage rates in the context of a 2001 population census for the town, rather than the one taken in 1991. By this calculation, cancer rates appeared lower than they really were. It is worth noting that a previous study, carried out by the chief medical officer of Somerset, concluded that there had been a significant increase in the incidence of leukemia and non-Hodgkin's lymphoma among people under the age of 25 since the Hinkley reactor went on stream in 1964.

Progress before people

The British nuclear industry has made repeated denials that living near a nuclear power station equates with an increased risk of cancer. Yet in February 2000, the nuclear installations inspectorate in the U.K. released damaging reports concerning the falsification of documents at the state-owned company British Nuclear Fuels (BNFL). The *Guardian* newspaper revealed that in March 2000, the director of the government environment agency had expressed concern about the apparent lack of commitment and responsibility that BNFL executives had for the environment. Furthermore, the minutes of a March 2000 meeting of the agency revealed that recent breaches of safety regulations at nuclear power plants in the U.K. had led to three enforcement notices against

Considerable uncertainty exists over the level of safety in storage facilities for nuclear waste. Evidence exists of leakage into groundwater.

26th APRIL 1986

REMEMBER CHERNOBYL

GREENPEACE

BNFL, the most serious of which was that of radioactive carbon-14 inadvertently discharged into the atmosphere from Hinkley.

In April 2002, the *Guardian* ran another article disclosing that, at the Sellafield nuclear reprocessing plant in Cumbria, radioactive material has been leaking from tanks containing massive quantities of untreated nuclear waste, and this has now contaminated groundwater in the area. In November 2001, borehole tests by the nuclear installations inspectorate found evidence of water contamination by a radioactive substance called technetium-99, which remains active for 200,000 years and can accumulate in the bodies of shellfish.

Also in November 2001, the British government accepted that BNFL was virtually bankrupt, with liabilities at Sellafield alone amounting to £34 billion ($54 billion). Yet in May 2002, the *Observer* newspaper reported that BNFL had recently spent more than £650,000 ($1 million) of taxpayers' money on donations to the U.S. Republican Party and on hiring White House lobbyists. The intention was to secure President Bush's approval to construct nuclear reactors in the U.S.

Greenpeace project a message onto Big Ben, at the British Houses of Parliament, to make sure the politicians and people of Britain do not forget what happened at Chernobyl and heed the warnings of the past.

Cows graze in a field adjacent to the Sellafield nuclear plant, Cumbria, U.K. But is the grass on which they feed contaminated?

REFERENCES

Barnett, A. and S. Hughes. BNFL spent $1 million lobbying in US. *The Observer.* May 19, 2002.
www.guardian.co.uk/nuclear/article/0,2763,718407,00.html

Brown, P. BNFL faces public safety "reminder." *The Guardian.* April 15, 2000.
www.guardian.co.uk/nuclear/article/0%2C2763%2C194427%2C00.html

Brown, P. "Cancer link" to nuclear plants. *The Guardian.* April 13, 2000.
www.guardian.co.uk/nuclear/article/0%2C3858%2C3985440%2C00.html

Brown, P. N-waste leak at Sellafield. *The Guardian.* April 18, 2002.
www.guardian.co.uk/nuclear /article/0,2763,686080,00.html

4.3 PCB Poisoning of North America

A dead fish provides striking evidence of contamination by PCBs and other chemicals in the Hudson River at Cold Stream, New York.

According to the Web site of the Department of Environmental Protection (DEP) in the U.S. state of Pennsylvania, its mission is "to protect the state's air, land and water from pollution, and promote a cleaner environment for health of citizens." A righteous claim, but in April 2001, Pennsylvania state representative and House Democratic Whip Mike Veon painted a rather different picture. He called on the acting secretary of the DEP to launch an investigation into the way it gave out health advice and urged action against "any individuals who may have acted intentionally to mislead the public by withholding scientific data."

Veon's challenge came after the Pennsylvania DEP made a dramatic U-turn over an earlier statement that several million stocked trout swimming in the state's waters were safe to eat. Suddenly, the department was warning that these same fish were carrying abnormally high levels of cancer-forming polychlorinated biphenyls (PCBs) and mercury that could cause fatal brain damage. According to Veon, the DEP "either hid its evidence or is so incompetent that it can't interpret the results... if the department is really this incompetent, then we need to set up a better system to protect the lives of Pennsylvanians and the resources in this state."

A widespread problem

This warning relating to public information and response over PCB pollution applies no less in various other parts of the United States and Canada. Nowhere has it been more applicable than in relation to the

Hudson River that runs southwest from the Adirondack Mountains and then east to the Hudson Falls, through New York state and finally south to the Atlantic. New York state sued the General Electric Company (GEC) for releasing PCBs into the river from electrical plants north of Albany from the 1940s to the 1970s. GEC argued that the state attorney general, Eliot Spitzer, had overstepped his authority and that the issues should be dealt with by the U.S. Environmental Protection Agency (EPA) and the state Department of Environmental Conservation (DEC). Thus began a long-running legal fight in which money and public profiles seemed more important than remedying the problems.

The chemicals giant Monsanto began PCB manufacture in 1929, and General Electric became a significant user of these compounds. By 1968, however, following a poisoning case involving PCBs in Yusho, Japan, U.S. authorities had become aware of the chemical's toxic effects. It wasn't until February 1976 that an administrative hearing of the DEC found that PCB contamination in the Hudson River was due to corporate abuse and regulatory failure, and Congress passed the Toxic Substances Control Act banning manufacture of PCBs. It was not until the following year, however, that Monsanto ceased production of PCBs, too late to avoid environmental consequences.

Passing the buck

When the extent of contamination in the Hudson River came to be known, complicated and protracted legal actions ensued over the cleanup and relocation of toxic waste to be removed by dredging. Much of this involved wrangling over who was to pay for the cleanup operation. In 1983, the upper Hudson River became earmarked as a national priority of Superfund, the federal program established in 1980 to investigate and clean up the most polluted sites in the United States. In September 1983, the EPA ruled that dredging the Hudson River did

Many Inuit people in Arctic Canada have PCB levels up to 70 times higher than Canadians who live further south.

not qualify for Superfund money and, at the same time, the federal court granted an indefinite extension of deadlines on release of so-called Section 116 money, the provision of county public funds for the job. In October 1984, the stalemate was resolved through lawsuits, and Section 116 money was finally released. It was not until December 1989 that the DEC announced its action plan to remove 250,000 pounds (113,400 kilograms) of PCBs from the Hudson River. In 1993, a settlement amounting to $7 million was reached between General Electric and commercial fishermen for lost income.

Meanwhile, the PCB pollution catastrophe goes on and contamination has spread alarmingly through the food chain. A paper published in 2000

General Electric reached a $7 million settlement in 1993 with fishermen whose incomes had been lost through contamination of the Hudson River.

revealed that Inuit people living in the Canadian Arctic carry PCB levels up to 70 times greater than people in the southern part of Canada. It is believed that this is due to a traditional diet containing high proportions of fat-laden meat from seals, whales, polar bears and other animals that, because they are near the top of the food chain and eat migrating fish, have themselves become heavily contaminated.

In Pennsylvania, state representative Mike Veon has made the somber observation that the condition of fish there serves as a warning mechanism that many U.S. waterways are more polluted than originally revealed. He, along with other public interest groups, has questioned the failure of Pennsylvania's DEC to advise the public and their alleged mistakes in the way the department has handled its data. It is difficult to avoid the conclusion that environmental authorities in more than one U.S. state, despite sanctimonious claims, have been reluctant to come clean about chemical pollution levels because of a possible conflict of interest—namely influential corporations that spend heavily in Washington and wield considerable influence in public policy, particularly on environmental issues.

REFERENCES

Courtney, D., C. D. Sandau, P. Ayotte, E. Dewailly, J. Duffe, and R. J. Nortstrom. 2000. Analysis of hydroxylated metabolites of PCBs (OH-PCBs) and other chlorinated phenolic compounds in whole blood from Canadian Inuit. *Environmental Health Perspectives 108:611-616.*
www.ourstolenfuture.org/NewScience/ubiquitous/2000courtney tal.htm

N.Y. sues G.E. over Hudson River contamination. *Court TV.* November 16, 1999.
www.courttv.com/national/1999/ge_suit_ap.html

PCB contamination of the Hudson River: A chronology. *Environmental Studies Program at Rensselear.* October 1994.
www.rpi.edu/dept/environ/orgs/Clearwater/chronology.html

Spitzer sues GE over Hudson River PCB contamination. *The Business Review (Albany).* November 15, 1999.
wwwamcity.com/albany/stories/1999/11/15/daily1.html

Veon wants investigation on state's reversal of PCB fish contamination. *Pennsylvania House Proceedings.* April 17, 2001.
www.pahouse.net/pr/veon/014041601.htm

William Aylign, from O'Brien and Gere Engineers Inc, pulls up a pail of water from the upper Hudson River. The engineering firm is monitoring the PCB levels on the Hudson.

4.4 The Love Canal

A local family cycling next to the contaminated area of the Love Canal in Niagara City, New York.

The saga of the inappropriately named Love Canal shows that the most obvious suspect is not always the real villain. Here was a case of toxic waste dumping on a massive scale, yet the industrial corporation to which the lethal leftovers belonged was largely above reproach in as much as it adopted a conscientious policy and did a great deal, albeit ineffectually, to dissuade the authorities from what followed. It was, in fact, the city politicians who were at fault in this instance.

From dream to disaster

The historical facts of the case are clear enough. Riding the bandwagon of American industrial expansion at the end of the nineteenth century, an entrepreneur named William T. Love came up with a scheme for a planned industrial community (incongruously known as "Model City") on the southern edge of Niagara City in upstate New York, which would have ready access to cheap hydroelectric power. Crucial to its success was the hydroelectric power that was to be provided by diverting some of the waters of the nearby river, famous for its spectacular falls, through a new urban canal. Love's dream gained approval, but although work began on the canal, his Model City never happened. The waterway became redundant and for some 50 years a vast trench punching north into the La Salle area of the city remained derelict until, in 1942, the Hooker Chemicals and Plastics Corporation bought it for use as a landfill site. During the next 11 years, the Love Canal received about 22,000 tons of mixed chemical waste, mainly pesticide, and chemical weapons material. By 1953, Hooker had filled the site to capacity and set about capping it off with a thick layer of hard-packed clay, using the most up-to-date techniques available at the time.

At this juncture, the Niagara City authorities were on the hunt for land to meet political pressures for urban expansion and badly needed housing. Hooker warned the city in the strongest language about the safety of the site, but the advice was ignored and the city forced its purchase on a constitutional technicality. As a mark of disapproval, the Hooker Corporation sold the site for $1, the minimum price for a binding contract, and added an unequivocal disclaimer warning of the danger. It did not, however, follow best engineering ethics by taking the city to court. Residential development went ahead, involving the laying of sewers that repeatedly penetrated the clay capping. The 99th Street Elementary School was the first building to go up on the site, opening its doors to pupils in 1955, after which the area was developed further and became a typical suburban neighborhood.

Conveniently forgotten hazards

Was this typical? Well, not quite, because there remained the matter of 22,000 tons of toxic industrial waste lying immediately beneath residents' properties. During the next 20 years, telltale signs began to emerge. There were unpleasant smells, unusual seepage into basements and reports of health problems. Little that was truly newsworthy happened, however, until 1975 when abnormally heavy rain and snow served to muddy the La Salle waters in more senses than one. Groundwater levels rose and part of the site subsided so that rusting metal drums began to appear and a cocktail of chemicals started to leak through the surface.

By August 1978, the evidence of dangerous chemical pollution was sufficiently obvious to prompt the New York Department of Health Commissioner, Robert Whalen, to close the 99th Street School, evacuate vulnerable residents, including pregnant women and children, fence off the area immediately around the landfill site and label it a threat to human health. In the following year, a scientist from the Roswell Park Memorial Institute in Buffalo, New York, was detailing a high rate of birth defects and miscarriages among residents. It was enough to persuade then President Jimmy Carter to declare a state of emergency, order federal funds and other disaster assistance to be placed at the disposal of the City of Niagara Falls and prompt a further temporary evacuation.

Residents of the houses built upon the Love Canal attend a public meeting.

One of Hooker Chemicals and Plastics Corporations factories, from which the chemical waste in Love Canal came.

Who was at fault?

This much is historical fact, but the ensuing claims and counterclaims amount to much less clear-cut evidence. The biggest problem, one in which the Love Canal incident by no means stands alone, has been the lack of readily available information that would allow people to make a properly informed judgment on one basic question: Is it safe?

In 1998, Dr. Elizabeth Whelan wrote an editorial for the American Council on Science and Health in which she rightly pointed out that by 1978 Love Canal had become a national media event with articles referring to the neighborhood as a "public health time bomb." Indeed, 248 different chemicals, including nearly 132 pounds (60 kilograms) of lethal dioxin—enough to kill millions of people—had been unearthed from the canal. Yet Whelan claimed that the only proven health problem was the stress induced by media hype, giving residents the impression that they and their children were going to become ill. She was not the first to voice such bizarre opinions. From 1980 onward, scientists and other Environmental Protection Agency officials had been asserting confidently that there was "no demonstration of acute health effects of hazardous waste exposure at Love Canal," while adding a prudent caveat that "as yet chronic effects of hazardous waste exposure have neither been established nor ruled out."

People were left wondering whom to believe and who was covering for who. The panel of scientists from the prestigious Memorial Sloan-Kettering Cancer Center, who had given safety assurances to the public, were the appointees of the New York State governor, Hugh Carey. How

In June 1979, U.S. President Jimmy Carter declared a state of emergency in La Salle, an area of Niagara City, and ordered disaster assistance.

In the area around the Love Canal, 700 families were evacuated so that the cleanup operation could begin.

could they be trusted to tell the truth? The authority that hired them had surely been aware back in the 1950s that a residential area was being constructed in Niagara Falls City, against corporate advice, on top of a vast toxic waste dump. And if there was no danger, why had the landfill site been reburied under a heavy plastic liner, clay and topsoil with 16 acres (6.5 hectares) surrounded by an 8-foot (2.4-meter) high barbed wire fence and declared permanently off-limits?

Today, the area of the Love Canal is curiously renamed Black Creek Village, though in an effort to polish up its tarnished image, there is a further proposal that it should adopt the quaintly sanitized title of "Sunrise City." The Occidental Corporation that evolved from Hooker Chemicals and Plastics has agreed to pay $230 million toward cleanup costs. Yet the real culprits are the city politicians who acted with total disregard of the dangers spelled out to them by Hooker, failing to inform residents about what they were literally sitting on, and the health department officials who, out of self-interest, then tried to cover up the mistakes for another 20 years.

REFERENCES

Essay for undergraduate engineering class. What happened at Love Canal? Undated.
http://cems.alfred.edu/students98/allansm/Onemoretry.html

Science and Engineering Library exhibit, University at Buffalo. Love Canal @ 20. Undated.
http://ublib.buffalo.edu/libraries/units/sel/exhibits/lovecanal.html

History of the Love Canal.
http://www.essential.org/orgs/cchw/lovcanal/lcsum.html

Key dates and events at Love Canal.
http://www.essential.org/orgs/cchw/lovcanal/lcdates.html

Gibbs, L.M. Learning from Love Canal: A 20th anniversary retrospective. Undated.
http://www.envirolink.org/enviroarts/arts_and_activism/Lois Gibbs.html

Whelan, E.M. 'Love Canal: Health Hype vs. Health Fact.' Editorial. American Council on Science and Health. December 26th 1999.
http://www.acsh.org/press/ed_archives99.html

4.5 The GM Food Debate

→

Greenpeace activists protest against genetically modified soy products in front of the Science and Technology Ministry in Brasilia in Brazil. Inside, a bio-security commission was discussing the regulation and commercialization of genetically-enhanced soy beans, which could be released into the marketplace.

The turmoil over whether or not genetically modified (GM) foods pose a threat to our health and represent the advent of a futuristic landscape populated with "Frankenstein crops" rages on, thick with insinuation. It becomes difficult to present an objective view without being labeled an ecowarrior or a stooge of the bioengineering commercial lobby, and it is clear that vested interests, both commercial and ideological, are being pursued vigorously from both sides. Yet through all the claim and counterclaim, it is possible to detect a serious coverup on the part of the pro-GM lobby, a reluctance to tell us, the feeding public, quite everything. Information trickling out in recent years in a plethora of news reports and litigations can be pieced together to shed light on some of the less well publicized areas of activity.

The seeds of dissent

One of the indications that all is not harmonious, even within the GM industry, has been the firing of Dr. Arpad Pusztai, the head of a research project at the Rowett Institute in Aberdeen, Scotland, after he raised the first major alarm over results of GM tests he had supervised. In August 1998, Pusztai disclosed details of experiments that had involved feeding laboratory rats for 100 days with genetically modified potatoes treated with similar strains developed by commercial food producers to make such crops pest-resistant. The laboratory animals had developed inflammation and thickening of their stomach lining, and Pusztai claimed that GM food could weaken immune systems and stunt growth while damaging the rats' internal organs. The response of the institute was to suspend Pusztai and to issue a statement: "The Institute regrets the release of misleading information about issues of such importance to the public and the scientific community."

A U.S. scientist wearing protective clothing sprays genetically modified strawberries, yet the long term consequences of such trials are the subject of considerable debate.

→

The difference in size between a natural strawberry and a genetically modified strawberry. The left fruit has had extra chromosomes (polyploidy) added.

French farmers, supported by a few hundred consumer group members and ecology activists, reap what they call an "illegal" plot of genetically modified rape seed, grown by U.S. company Monsanto near Lyon. The demonstrators delivered their harvest to the company's offices in the suburbs of Lyon.

Preventing justice

Pusztai then submitted his findings and those of a colleague, Dr. Stanley Ewen, to the British medical journal the *Lancet*, and in spite of heavy objection from referees, claiming the research was "unsound," the article was published in October 1999. The scathing and emotive rhetoric adopted by some of the referees might be construed as an indication that they were privately more troubled by Pusztai's findings than they were prepared to admit. Professor John Pickett, for example, asserted vehemently, "If this work had been part of a student's study, then the student would have failed whatever examination he was contributing the work for." Pickett, however, was an employee of the Institute of Arable Crops Research (IACR) in Hertfordshire, U.K., an organization sponsored by the Biotechnology and Biological Sciences Research Council.

In March 2001, a U.S.-based Web site, hosted by Dr. Joseph Mercola, published an article entitled "Suppressing Dissent in Science with GM Foods." The contributors, Dr. Mae-Wan Ho and Jonathan Matthews, stated that the U.K. Royal Society had set up a review of Pusztai's results without giving him the opportunity to submit the complete data files and then published their own report declaring his findings flawed. The article also reported that dependence on private funding is critical in Britain's top research universities, often amounting to 80 to 90 percent of the total research budget. It points to a survey of scientists working in government or in recently privatized laboratories that reveals worrying statistics. One third of those polled indicated that they had been asked to alter research findings to suit the customer, while 10 percent had been pressured to massage results.

Bias research

These allegations must be considered in light of a February 2000 BBC News report that U.K. biotechnology companies were funding a scientific panel, known as Cropgen, with a budget of just under £500,000 ($800,000) "to help achieve a more balanced debate about GM crops." The sponsors have agreed that they will not veto any of the scientific positions taken by the panel, but Cropgen does not speak with a neutral voice. Three of the eight panel members work, or have worked, for the IACR organization Professor Pickett has also been associated with, and the initiative for the program stemmed from biotechnology companies. The panel chairman, Professor Vivian Moses, has admitted that its mission "is to provide a voice for crop biotechnology—a voice that has all too often been missing from the public debates in the U.K. to date." Set beside these sentiments, research published by the Centre for Food Policy has recently criticized the British government over the fact that only 10 percent of its research projects on GM food investigate the safety issue; the rest focus on the best methods of commercially exploiting GM technology.

Also in February 2000, the *Daily Mail* newspaper reported no less worrying evidence that pro-GM interest groups have contributed to a virtual coverup, and that the release of GM foods for public consumption has been based on flawed experiments that have failed to confirm their safety. Scientists with the U.S. Food and Drug Administration (FDA) fed GM tomatoes to rats in a 1993 study, but could not determine from the tests that GM products are harmless. The study revealed problems similar to those detected by Pusztai in 1998. Yet concerns raised by these trials appear to have been hushed up. Worse, they effectively form the scientific justification for the release of a range of GM products, including soya and corn.

The *Daily Mail* article quotes an Iowa-based attorney, Stephen Druker, as having accused the FDA of "misrepresentations that are not innocent, they are fraudulent...the agency's behavior is not only illegal and irresponsible, it is unconscionable." Druker's comments came in the

An environmental activist dressed as the grim reaper demonstrates in a field of genetically modified maize.

context of a legal action by U.S. farmers and GM critics against Monsanto and other biotech companies. Evidence submitted to the court confirmed that Pusztai had not spoken with a lone voice when files were produced containing "memorandum after memorandum" from FDA technical experts warning about the potential risks of genetically engineered foods.

The *Daily Mail* article also reported that while biotech companies and the authorities in the U.K. and the U.S. were insisting that human feeding trials with GM foods were unnecessary because GM foods are "substantially equivalent to natural products," other documents clearly stated that "GM foods cannot be presumed to be substantially equivalent to conventional foods and that they entail a unique set of risks."

Tests on rats fed with genetically modified tomatoes like these failed to determine that GM products are harmless. Some research findings are alleged to have been doctored.

REFERENCES

Barnett, A. GM genes "jump species barrier." *The Observer.* May 28, 2000.
www.guardian.co.uk/Print/0,3858,4023082,00.html

Blythman, J., R. Edwards, and P. Taylor. U.S. covered up GM food fears. *The Sunday Herald.* February 27, 2000

Brown, P. Lawyer's challenge to U.S. over GM safety claims. *The Guardian.* February 29, 2000.
www.guardian.co.uk/Print/0,3858,3968705,00.html.

GM controversy intensifies. *BBC News Online.* October 15, 1999.
http://news.bbc.co.uk/hi/english/sci/tech/newsid_474000/474911.stm

Ho, Mae-Wan and J. Matthews. Suppressing dissent in science with GM foods. *Optimal Centre.* March 14, 2001

Kirby, A. GM firms fund friendly scientists. *BBC News Online.* February 25, 2000.
http://news.bbc.co.uk/hi/english/sci/tech/newsid_655000/655479.stm

Lancet defies GM study advice. *BBC News Online.* October 15, 1999.
www.news.bbc.co.uk/hi/english/sci/tech/newsid_472000/472192.stm

Meikle, J. Soya gene find fuels doubts on GM crops. *The Guardian.* May 31, 2000.
www.guardian.co.uk/Print/0,3858,4023886,00.html

Pusztai "to be vindicated." *The Guardian.* October 4, 1999.
www.guardian.co.uk/Archive/Article/0,4273,3909029,00.html

Secret papers show scientists are at odds over risks. *The Daily Mail.* February 11, 2000

5 A Green and Red Herring

The separate truths of governments and greens

5   A Green and Red Herring

It would be unfair to judge all scientific secrecy as being a phenomenon that arises from corporate and government conspiracy and greed for money. Sometimes, coverups can be attributed to other, arguably more altruistic, motives.

↑

Deforestation for timber production continues in many tropical forests, but how do we know which of the conflicting arguments to believe?

Ideology can play a strong part, particularly in instances where data can be interpreted in different ways. We, the general public, are not afforded much of the information on which a case can be judged and even if we were, it is probable that we would not possess the expertise to evaluate the details effectively. So we put our trust in others. We rely on those whom we believe to be the honest brokers of the secrecy game, but our trust also tends to be formed subjectively. When we are given a choice between believing the words of fat industrial conglomerates and those offered to us by a pressure group or a government agency that is supposedly on our side, there is an instinctive reaction to put our money on the latter. But that is not always the wise choice.

Nowhere is secrecy for ideological motives demonstrated more sharply than when information is exploited selectively by pressure groups. In much of the debate about such issues as global warming, deforestation and acid rain damage, conflicting scientific figures and analyses are bandied backward and forward, leaving us somewhat dizzy and not much the wiser. Evidence emerges suggesting that groups like Greenpeace and Friends of the Earth are just as capable of massaging data to suit their own arguments as the big industrial corporations. It would be reasonable to assume that the truth lies somewhere in the middle of the two arguments we are being given, but our beliefs should be based on fact, not supposition and interpretation. How are we to solve the potentially catastrophic environmental problems of the world without definitive information?

Governments are not averse to employing diversion tactics either. The Norwegian whaling issue provides a trenchant illustration. The International Whaling Commission broadcasts one set of figures warning us that Minke whale stocks in the North Sea are being dramatically depleted and that the animals are being placed at serious risk from overfishing. On the other hand, the Norwegian government claims that the same Minke whale stocks are in no danger and that commercial whaling is acceptable. We need to keep a clear head and remember that one organization is interested in maximizing the case for animal welfare, while the other is anxious to satisfy the needs of local communities that are dependent on whaling and that, incidentally, vote to keep the Norwegian government in power. How do we gauge which of the two is telling us the truth? After all, few of us could afford to spend a year surveying the North Sea to assess the size of its whale stock even if we had the technical means of doing so.

It is clear what the motivations are for pressure groups and environmental bodies, but what are they for governments? Just like in so many aspects of politics, it is based on economics and international relations. It is convenient for the western powers that two-thirds of the world's nuclear waste is stored in northwest Russia, a thinly inhabited area that is well away from the prying eyes of most of the world's media. It allows governments, on the whole, to deny the problems that arise from spent nuclear fuel rods from power stations and nuclear submarines. Similar reasoning lies behind the secrecy over the ever-increasing space debris that circles the earth. "Out of sight, out of mind" is an attitude adopted in many instances of political and environmental refutation. Just how much damage will be done to the Earth before solutions are sought?

Carbon dioxide emissions from industrial plants have been labeled as the culprits in global warming.

As the world has become more consumer-led, it has also become more heavily populated. Maintaining the current lifestyles of billions of people puts pressure on the environment in dozens of ways.

5.1 Global Warming

→

Scientists from the University of Gottingen, Germany, studying climate change.

If there is any single environmental issue that has attracted the attention of the public in the last decade, it is the prospect of our planet getting progressively warmer through the adverse effects of greenhouse gases building up in the atmosphere. Carbon dioxide emissions from industrial plants and automobile exhausts around the world are labeled as the culprits and, according to some scaremongers, unless we reduce industrial output drastically and use bicycles to get around, the Earth is doomed to suffocation. The rejection by the United States of the Kyoto agreement on reduction of greenhouse gases, concluded in Japan in December 1997, brought the whole issue into sharp focus. Yet the issue of global warming is saddled with confusion and contradiction.

There's no such thing as objective

News coverage of Kyoto appears to have been heavily slanted in the U.S. press. According to the Media Research Center, no more than two of the 48 Kyoto-based reports on U.S. network news presented the arguments of dissenting scientists and only nine mentioned that disagreement actually existed. A news release of April 1998 from the American Science & Environmental Policy Project (SEPP) also claims that White House assertions that the theories of the Intergovernmental Panel on Climate

| Mar. 1979 | Mar. 1980 | Mar. 1981 | Mar. 1982 |

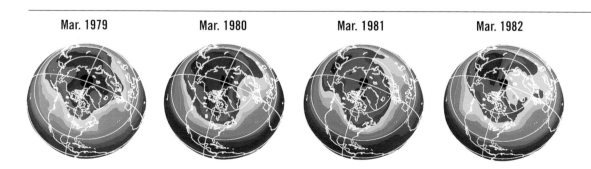

Change (IPCC) reflected a "consensus of 2,500 climate scientists" are a complete fabrication. The news release indicates that the scientists actually numbered less than 2,000 and only a fraction are said to be climatologists.

Scientific facts appear to have been covered up, distorted or flatly contradicted and the blame cannot necessarily be dropped onto the shoulders of multinational corporations seeking to protect their commercial interests. Not least among the parties guilty of hush-hush in this instance are the greens and ecowarriors determined to press home the ideological message that we would all be better off abandoning capitalism and reverting to a simplified way of life without many of the comforts of modern living. This ideology, it has been argued, has resulted in the misuse of science to promote a political agenda.

The final plenary session of the United Nations global warming conference in Kyoto in 1997. A treaty aimed at heading off a potentially catastrophic warming of the Earth was formally adopted after 11 days of talks, although not all nations signed the treaty.

Demands for proof

In April 1998, another voice came to the fore. No less than 17,000 scientists, rather more than one could imagine being coerced by big business, signed the Oregon Petition, which was prepared by the Oregon Institute of Medicine and Science. In submitting it to the U.S. government, they expressed their profound skepticism about the science underpinning the Kyoto agreement and urged its rejection. Many of those who added their names pointed to flawed computer models predicting future climatic trends and some mentioned that global warming is part of the planet's natural cycle. In an accompanying letter, Dr. Frederick Seitz, an emeritus professor at Rockefeller University and a former president of the U.S. National Academy of Sciences, stated that "the treaty is, in our opinion, based on flawed ideas. Research data on climate change do not show that human use of hydrocarbons is harmful. To the contrary, there is good evidence that increased atmospheric carbon dioxide is environmentally helpful."

Ozone mapping spectrometers reveal reductions in levels of the protective gas in the northern hemisphere 1979–1994. In the Antarctic, however, the hole in the ozone layer is now known to be closing due to reduction in worldwide use of the CFC productions.

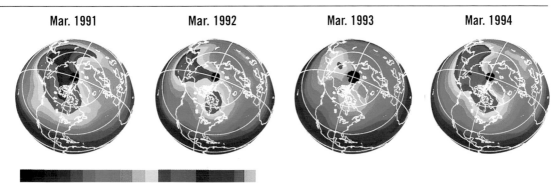

| Mar. 1991 | Mar. 1992 | Mar. 1993 | Mar. 1994 |

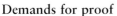

200.   **Total DU**   600.

The change in temperature in the earth's atmosphere has been blamed for the melting of icebergs, rising water levels and freak flooding. This girl was caught in high tides on the island of Tuvalu, the world's smallest country.

The UN-sponsored IPCC based much of the argument in favor of the global warming theory on a 1996 report that has been hailed by the mass media and environmental pressure groups as the authoritative statement on the issue. Yet according to Seitz, a comparison of the report that was approved by the scientists and the published final version reveals significant changes that were made after the scientists had met and accepted what they thought was to be the final version. Among passages deleted from the approved report are a number of crucial statements, including the following:

None of the studies cited above has shown clear evidence that we can attribute the observed [climate] changes to the specific cause of increases in greenhouse gases. No study to date has positively attributed all or part [of the climate change observed to date] to anthropogenic [man-made] causes. Any claims of positive detection of significant climate change are likely to remain controversial until uncertainties in the total natural variability of the climate system are reduced.

Seitz stated angrily that he had "never witnessed a more disturbing corruption of the peer review process than the events that led to this IPCC report." He made the strong accusation that "whatever the intent was of those who made these significant changes, their effect is to deceive policy makers and the public into believing that the scientific evidence shows human activities are causing global warming."

The Oregon Petition, it is said, reflects the frustration of the scientific community's "silent majority" over the hype emanating from politicians and much of the media about a "warming catastrophe." It is worth noting that the petition's funding came from voluntary donations, with no contributions from industry. SEPP President Dr. Fred Singer has spoken of scientists being appalled at the amount of research funding, worth $2 billion a year, which is diverted into "community workshops" that are little more than brainwashing exercises to enhance fears about climate change.

REFERENCES

Global warming petition project. *Oregon Institute of Science and Medicine.* 2001.
www.oism.org/pproject/s33p37.htm

Knautx, R., ed. The global warming debate. *Free Market Net.* February 1998.
www.free-market.net/features/spotlight/9802.html

More than 15,000 scientists protest Kyoto accord; speak out against global warming myth. *The Science and Environmental Policy Project.* April 20, 1998
www.sepp.org/pressrel/petition.html

Seitz, F. A major deception on global warming. *Wall Street Journal.* June 12, 1996.
www.sepp.org/glwarm/majordeception.html

5.2 The Silencing of Bjorn Lomborg

In October 2001, Cambridge University Press published a controversial book, *The Skeptical Environmentalist*, by Danish associate university professor and statistician Bjorn Lomborg. He questioned what he describes as the "litany of our ever-deteriorating environment," which is paraded by the green lobby. Using detailed figures and examples drawn from research into environmental statistics over the past 30 years, he threw considerable doubt on the accuracy of conventional arguments about the extinction of species, pollution of the air we breathe and the water we drink, deforestation due to acid rain and so on.

Woodland in the Czech Reupblic devastated by the supposed effects of acid rain.

Establishment on the defensive

What happened next was remarkable. Almost immediately on publication, the book and its author ran into howls of protest that at times have verged upon apoplexy. Some environmentalists called Lomborg a liar and a fraud while others refused even to share a public platform with him. The eminent journal *Scientific American* published an 11-page critique compiled by a group of scientists known for their support of the green movement who set out to trash his opinions. Even the journal's editor trumpeted, "Science defends itself against the *Skeptical Environmentalist*."

Yet careful examination of the critique reveals that while it is strong on rhetoric, it appears equally weak on substance. The reaction of

Flames from burning oil wells in Kuwait following the 1991 Gulf War, which took firefighters almost a year to extinguish. In addition to the massive oil spills, the smoke from the fires caused immense damage to the local environment. The smoke also contributed to short-term local climate changes, but did not significantly affect global climate systems as had been feared.

Professor Stuart Pimm, a conservation biologist at Columbia University, is typical of the establishment invective. He resorts to a well-trodden path followed by academics out to denigrate those whose views risk upsetting the favored applecart. "He [Lomborg] occupies a very junior position in Denmark...he has no publications on environmental issues, and yet manages to dismiss the science of dozens of the world's best scientists. As any sensible person would expect, his facts are usually fallacies and his analysis is largely nonexistent." One might be forgiven, as the editor of the *Economist* put it in January 2002, for thinking that Pimm and others sharing his views protest too much and are actually more rattled because the book knocks a powerful and convincing hole in the environmental movement's case. When Lomborg chose to publish the hostile review on his own Web site along, with a defense against his critics, *Scientific American* threatened legal action if he did not withdraw the text of their critique.

Lomborg's investigation is thorough, based on widely accepted UN data, and he backs his argument with meticulously detailed examples. Yet his diligence in providing references from literature is seized on by one of the *Scientific American* critics as "affectation." The hostility of the anti-Lomborg camp may, however, have less to do with defending honest scientific principles and countering an argument (which they seem largely unable to do) than with a more devious scenario. The *Economist* has pointed to an interview given in 1989 to the award-winning science and technology magazine *Discover* by Stephen Schneider, one of the coauthors of the *Scientific American* critique. While explaining the necessity to capture the public's imagination with substantial media coverage, Schneider came out with a particularly revealing strategy. "We have to offer up scary scenarios, make simplified, dramatic statements, and make little mention of any doubts we might have...Each of us has to decide what the right balance is between being effective and being honest."

The two extremes

Environmental activists seem concerned, by fair means or foul, with trashing any suggestions that we may not be on the brink of their impassioned scenario: environmental apocalypse. The green lobby is, for example, fond of claiming evidence of catastrophic deforestation through the effects of acid rain. In 1996, however, the U.S. National Acid Precipitation Assessment Program (NAPAP) concluded quite the opposite, that "most forest ecosystems are not currently known to be adversely impacted by acid deposition." Even trends in the right direction do not necessarily suit the political arguments of environmentalists. In 2000, the prestigious United Nations Food and Agriculture Organization (FAO) published its latest Forest Resources Assessment, which revealed that the global rate of forest loss is now 20 percent lower than the figure previously reported in 1995. The response of leading environmental pressure groups was to claim that the analysis was "fatally flawed."

In December 2001, a correspondent to *Nature* magazine noted that Lomborg's book brings into the open the continuing disparity between, on the one hand, disaster claims over the future of mankind and, on the other, evidence from large organizations such as FAO that offer the opposite. Another commentator made the succinct observation that the refusal of some environmentalists to deal honestly with the data in pursuit of their goals actually harms their credibility.

We should, perhaps, think twice before always accepting at face value conventional thinking that bows to popular ideology. The ferocity of the response against Bjorn Lomborg prompts the question of whether the vehemence raised against him has something to do with his focus on basic flaws in the political dogma of the greens. What he has done, in certain respects, is to expose the truth that environmental activists are as capable of tailoring the facts when it suits their purpose as are the leaders of industry and commerce.

View of a cloud of atmospheric pollution almost obscuring the sun and causing the sun's light to turn red.

REFERENCES

Budiansky, S. Diversionary tactics in environmental debate. *Correspondence in Nature.* January 24, 2002.

Defending science (editorial). *The Economist.* January 31, 2002.
www.economist.com/displayStory.cfm?Story_ID=965718

Gilland, T. A statistician with a mission. *Spiked Science.* August 30, 2001.
www.spiked-online.com/articles/00000002D214.htm

The litany and the heretic. *The Economist.* January 31, 2002.
www.economist.com/displayStory.cfm?Story_ID=965520

National Acid Precipitation Assessment Program. *NOAA Research.* January 20, 2000.
www.oar.noaa.gov/organization/napap.html.

Reynolds, G. H. Free speech under attack. *TCS.* Undated.
www.techcentralstation.com/1051/envirwrapper.jsp

The Skeptical Environmentalist. *Cambridge University Press.* Undated
http://uk.cambridge.org/economics/lomborg/

Trewavas, A. Open debate is essential on conservation issues. *Correspondence in Nature.* December 6, 2001.

The truth about the environment. *The Economist.* August 2, 2001.
www.ocs.orst.edu/forum/Lomborg2.html

5.3 Sweden's Biggest Environmental Disaster

Skanska chief Jan Stattin inspecting the tunnel walls.

About 62 miles (100 kilometers) north of the city of Malmö in southern Sweden, a small mountain extending east and west, and 985 feet (300 meters) in height, rears up from the plain at Hallandsas, near the town of Bastad. Consisting largely of sand, cracked rock and clay containing large amounts of water, it stands in the way of a section of the route that will provide one of the "missing links" in Scandinavia's rail network. It also represents an environmental disaster in the form of loss of groundwater from the region and contamination by the polymer acrylamide.

The need for progress

A Swedish SJ international train that would have been travelling through the Hallandsas tunnel had the project been successful.

For some decades, the railway system in Sweden, Banverket, had been in a state of decline and it was considered politically important that the country modernize its high-speed rail connections with the rest of Europe. Between 1985 and 1990, various national and local authorities launched studies intended to resolve the problem of how to take a new double track around or through Hallandsas. The Swedish government finally accepted a scheme for a two-tier, 5.3-mile (8.6-kilometer) railway tunnel. The Swedish Railway Administration was given the go-ahead for the project in 1991, with work commencing in the following year.

Problem upon problem

Problems arose almost immediately for the first contractor, Kraftbyggarna, because of the material consistency of the hill. The massive Hallbor boring machine, imported in 1993 from its previous employment in excavating the Channel Tunnel, ground to a halt after only a few yards, bogged down in clay and mud, and had to be withdrawn. In 1995, a new contractor, Skanska, took over, tunneling from three separate starting points. It rapidly became clear that an effective lining would have to be inserted because groundwater was now leaking into the tunnel at a substantial rate. A laboratory, contracted by Skanska, proposed a product known as Rhoca-Gil, manufactured by Rhodia, a subsidiary of the French multinational Rhone-Poulenc. Based on the polymer acrylamide, Rhoca-Gil has been used in various difficult construction projects worldwide and was said to be efficient as a sealant. And so, from the late spring of 1997, 1,400 metric tons of Rhoca-Gil were sprayed inside the excavated parts of the tunnel.

A water sample from inside the mountain is being tested by a worker in protective clothing.

Keeping quiet

It is alleged, however, that Skanska did not properly inform its workers or the local population about the level of toxicity of acrylamide, which is considered to be a dangerous human carcinogen and nerve poison. Information on the toxicity factor was readily available from at least the summer of 1994 through literature and Material Safety Data Sheets. Laboratory safety instructions prepared by the U.S. National Academy of Sciences in Washington, D.C., and other organizations, warned that in the short term the effects of contact with the compound include reddening and peeling of the skin, and eye irritation. The effects of longer-term exposure include unsteadiness, muscle weakness and numbness in limbs. Tests on animals suggest that acrylamide is cancer-forming.

In October 1997, some 6,000 people staged a protest march in Bastad against the Swedish Railway Administration. They were reportedly angered by reports that a fish-breeder on Hallandsas Mountain was finding dead fish in a stream and in ponds, and that cows in the neighborhood were becoming sick and lame. Work on the tunnel was stopped and it was then learned that several railway construction workers had shown adverse symptoms attributed to coming into contact with acrylamide. Apparently, the sealant had not set properly and acrylamide had dissolved into the groundwater of the hill, much of which was by then pouring into the tunnel.

From 1997, all work on the Hallandsas tunnel was stopped apart from inserting concrete lining into certain sections. The contamination resulted in the slaughter and burning of cattle and a ban on the sale of vegetables, meat and dairy products from the area. Subsequently, Skanska, Rhone-Poulenc and the Banverket all faced criminal investigations for violating worker safety laws. Skanska and the Banverket argued that they were unaware of the levels of acrylamide in the sealant. In November 1997, the Swedish Chemical Inspectorate brought criminal charges against Rhone-Poulenc Sweden for giving false information on its labels. Analysis by the commission of inquiry indicates, however, that neither the Banverket nor Skanska tried particularly hard to learn about Rhoca-Gil given that the hazards of acrylamide to human and animal health had been well documented.

In the autumn of 1997, in a separate incident, Rhoca-Gil was also used to seal the lining of a tunnel in Norway. The sealant again failed to set properly and acrylamide leaked into the drinking water system, leading to a serious pollution accident near Oslo. A fine of 3.5 million Norwegian kroner ($470,000) was served on Rhodia for failing to provide correct data sheets with Rhoca-Gil. The building contractor, NSB Gardermoen (Oslo), was also fined for failing to inform subcontractors of the level of toxicity in the product. Rhodia immediately suspended production of Rhoca-Gil and, although in October 1998 it was considering an appeal, it appears from Rhodia

Protest against the building of the tunnel. Opposition voices are now raised once again as the prospect of renewed drilling looms closer.

advertising that the company no longer manufactures Rhoca-Gil.

Effects on people and planet

The loss of water from the hill in Sweden has badly affected people living there. At the time that the original contractor abandoned the project, water was leaking into the tunnel at a rate of 15 gallons (58 liters) per second and the County Administrative Board in Kristianstad were demanding that the Banverket stem the leaks. According to a pressure group based in Bastad, Skanska allowed substantially higher quantities of water to flow from the tunnel than were permitted by the Swedish Water Court. The permitted extraction rate was 8.7 gallons (33 liters) per second, whereas up to 26 gallons (100 liters) per second were leaking away. As a result, more than 150 farms have now lost their water supply. The most recent proposal, however, is to pump water at 26 gallons (100 liters) per second for up to 10 years until the project is complete. This, say local environmentalists, will dry out the entire mountain. Recently, the Swedish Railway Administration was fined 3 million Swedish kronor ($325,000), the highest penalty possible in a Swedish environmental prosecution, for having breached the permitted water removal from the hill between 1992 and 1997.

Hunters were hired to shoot the sick and poisoned cows.

REFERENCES

Almqvist, J. Local anti-tunnel campaigner residing in Bastad

Company profile. *Rhodia*. 1999.
www.us.rhodia.com/profile/co_profile.htm

HHMI Lab Safety: LCSS Acrylamide. *National Academy of Sciences*. 1995.
www.hhmi.org/research/labsafe/lcss/lcss7.html

Karlsson, I. Visst gar det att bygga tunnlar I Halland. *Goteborgs-Posten*. March 11, 1998

Lofstedt, R. Off track in Sweden. *Environment*. May, 1999.
www.findarticles.com/cf_)/m1076/4_41/54711397/p1/article.jhtml

Rhoca-Gil. Skanska. Undated.
http://thehub.skanska.com/700_External/70002_Pressrelease.asp ?WebID=1&EntryID=2992&LangID=1;

http://thehub.skanska.com/700_External/70002_Pressrelease.asp ?WebID=1&EntryID=2091&LangID=1;

http://thehub.skanska.com/700_External/70002_Pressrelease.asp ?WebID=1&EntryID=1884&LangID=1

Scott, A. Norway fines Rhodia for acrylamide leak. *Chemical Week*. October 28, 1998.
www.findarticles.com/cf_)/m3066/1998_Oct_28/53132519/ print.jhtml

5.4 Deforestation

The general public is often bombarded with conflicting information about matters of universal concern. More often than not, it becomes fairly clear who is telling the truth and who is attempting a coverup. But on occasion, matters are not so obvious. Scientific data can be interpreted in different ways and nowhere is it more difficult to arrive at what is actually happening than in the impassioned case of the vanishing rain forests.

From one extreme...

On August 8, 2000, the United Nations Food and Agriculture Organization (FAO) published a report, based on analysis of more than 300 satellite images, indicating that tropical deforestation appears to have slowed by some 10 percent compared with the rate estimated during the 1980s. This document was a follow-up to a 200-page report released in 1997 by the FAO's Committee on Forestry (COFO). Between 1980 and 1990, the loss of natural forest in all developing countries was estimated to be 38.3 million acres (15.5 million hectares) per year. In the new findings, this deforestation figure decreased to 33.9 million acres (13.7 million) hectares annually during the 1990s.

...to the other

Notwithstanding the FAO warning that its most recent results do not indicate that the battle against deforestation is over, and that reduction in deforestation should not become an excuse for unsustainable forest practices, this would appear to be a positive trend. Yet this latest report, part of the FAO's Global Forest Resources Assessment 2000, received almost immediate howls of protest from environmental pressure groups.

The destruction of forests in Madagascar has had a profoundly damaging effect not only on the local economy but also on wildlife, much of which is unique to the island.

The response of Bruce Cabarle, director of the Global Forest Program at the U.S. division of the World Wildlife Fund for Nature (WWF) is not untypical. "WWF does not believe that deforestation is slowing down, but rather has continued at the same or even higher levels than in the 1980s, and that this is a cause for alarm rather than complacency."

Not knowing what to believe

One of the first indications of the difficulties in reaching any kind of accurate evaluation was an assessment of rain forest loss made in 1980. The study, *Conversion of Tropical Moist Forests*, was published by the National Academy of Sciences in the United States and was taken by environmentalists to be a benchmark. It calculated that between 50 million and 60 million acres (20 million and 24 million hectares) of closed, tropical broadleaf forest were being lost annually between 1976 and 1980. These figures proved of great influence to public perceptions.

The statistics were also used to project future deforestation rates in the much-publicized Global 2000 Report. The author of the study, Professor Norman Myers, a consultant in environment and development, had worked since 1970 in the general subject area of environment and natural resources, with emphasis on species, gene reservoirs and tropical forests. Yet Myers' estimates quickly proved unreliable and he was among the first to acknowledge that they amounted to "a crude approximation at best."

As early as 1982, data indicated that Myers' figures were vastly overestimated and that the annual loss of closed tropical forest was closer to 17.5 million acres (7.1 million hectares) a year. Whereas the 1980 report predicted annual losses of between 2.3 and 4.8 percent of

Not all deforestation can be blamed on logging activities. Here, in the Amazon basin, destruction has resulted from illegal cassitorite (tin ore) mining.

In 1980, the U.S. National Academy of Sciences estimated that up to 24 million hectares of tropical broadleaf forest was being lost annually through logging. Yet how accurate were their statistics?

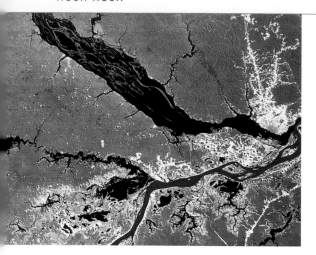

The Rio Negro River (predominant dark color river from upper left) can be seen joining the Rio Solimoes River (blue from the bottom) to form the Amazon River at Manaus. Areas of tropical forest appear as dark orange colors; areas of deforestation appear as blue to blue-green; and areas of regrowth appear as lighter orange.

the world's rain forests, subsequent and more accurate estimates using satellite imaging have pointed to a level of between 1.5 and 2 percent. In 1998, Brazilian authorities estimated that the country was cutting down about 2 percent of its rain forest each year, but a year later studies amended this figure to just 0.5 percent. The FAO put forward an estimate in the 1980s of deforestation in tropical rain forests amounting to 0.8 percent a year. In the 1990s, this was reduced to 0.7 percent and is now down to 0.46 percent.

Whether the most recent FAO report presents an accurate assessment or not is impossible for the average reader to evaluate, but as Anthony Janetos, senior vice president of the Washington, D.C.-based World Resources Institute (WRI), has conceded, "No other organization provides such comprehensive information on global forests as FAO." So why are its findings condemned in certain quarters? An astute observation to be found among the working papers of the European parliament indicates that the actions of nongovernmental organizations like the WWF, Greenpeace and Friends of the Earth have been aimed at "sensitizing public opinion to deforestation or forest degradation."

Emily Matthews, the author of a study for WRI, is one of those who joined in the clamor, arguing that "for FAO to say that global deforestation is slowing down is misleading given the differences in regional and subregional conditions of the world's forests." Yet Matthews' complaint lacks conviction, particularly since her argument can be cut both ways. The FAO report may equally overestimate rates of global deforestation. As Matthews points out, an obstacle to any accurate assessment of worldwide trends is that figures do indeed vary considerably among countries. In 1980, for example, Myers estimated the deforestation rate on the Ivory Coast of West Africa to be 5.3 percent but that of Brazil was only 0.342 percent.

The 2000 FAO analysis, it has been claimed, is "fatally flawed." Yet this condemnation has become almost the standard rhetoric used by any pressure group that wants to dissuade us from believing in the findings of others. Sometimes we, the public, need to keep our minds open over exactly who is telling the truth and in issues like that of deforestation, where data can be juggled to suit, understand that no one actually knows the truth. The experts can be guaranteed to work their calculations on different criteria, and most of us have neither access to sophisticated statistical information nor the expertise to understand it. Our lack of firsthand knowledge and experience may be exploited and we may be handed information that sounds highly convincing but is designed to mislead us for reasons of ideological gain and publicity. The maxim of *caveat emptor* or "let the buyer beware" is one we should remember.

Agricultural demands can also result in massive loss of forest. Here in the Amazon, trees are engulfed in flames in order to clear more land for agriculture.

REFERENCES

Allegations that new deforestation estimates are flawed. *Worldwide Forest/Biodiversity Campaign News*. December 3, 2001.
http://forests.org/recent/2001/wristrep.htm

European Parliament working papers. *Europe and the forest-Vol. 3. Europarl.* Undated.
www.europarl.eu.int/workingpapers/forest/eurfo236_en.htm

Forestry committee reviews state of the world's forests. *Food and Agriculture Organization of the United Nations.* 1997.
www.fao.org/News/1997/970301-e.htm

An improved method for monitoring national and global deforestation. *Science Daily.* June 1, 2001. Adapted from a news release posted by NASA/Goddard Space Flight Center. May 6, 2001.
www.sciencedaily.com/releases/2001/06/010605072649.htm

Sedjo, R.A. and M. Clawson. 1983. How serious is tropical deforestation? *Journal of Forestry 81:* 792-93.
www.ciesin.org/docs/002-112/002-112.html

Strong indications for slow-down in deforestation. *Food and Agriculture Organization of the United Nations.* August 8, 2000.
www.fao.org/WAICENT/OIS/PRESS_NE/PRESSENG/2000/pren0045.htm

5.5 The Minke Whale Deception

Greenpeace intervene as Norway continues to needlessly slaughter Minke whales.

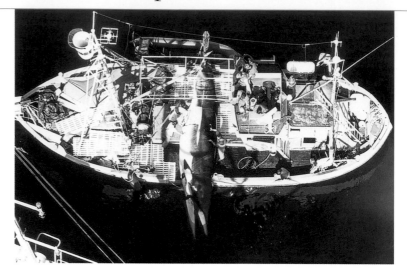

In 1985, an important meeting of the International Whaling Commission (IWC) took place at which, on the strength of mounting scientific evidence of threats to its survival, the Minke whale was labeled a "Protected Stock." Not everyone agreed, however, and the Norwegian representatives filed an objection, declaring that their country would not be bound by the ruling. For some years, the Norwegians played along, but in the summer of 1993, they shocked the IWC with a unilateral resumption of commercial hunting, claiming that the commission had used delaying tactics to re-evaluate the suspension on Minke whaling and therefore on discussing future catch quotas. But the issue was far less clear-cut. Despite Norway's righteous protestation that its whaling was environmentally sound—a harmless traditional industry catching food for local consumption—subsequent investigation revealed a catalog of cheating, dishonesty, violation of regulations and reliance on flawed data.

The whalers and their catch

The Minke (Balaenoptera acutorostrata) is the smallest species of baleen whale, a marine mammal found in all the world's oceans that feeds by filtering plankton from seawater. In 1991, the Scientific Committee of the IWC produced a disturbing report that the Minke population in the northeast Atlantic, which had hitherto been regarded as well stocked, had suffered a significant decline over a 30-year period, which was in part attributed to the Norwegian whaling fleet. Since the 1930s, these boats had been operating in the seas around the Lofoten Islands, but as the whales became scarce there, the boats were traveling further away.

During the 1950s, Norway had an annual catch of as many as 4,000 Minke, although this level was subsequently cut back. Norwegian whale hunters, it was claimed by conservationists, had enjoyed virtual freedom of operation for decades. Regulations were inadequate and those supposedly in force were regularly violated and poorly policed. No actual catch quotas had been set until 1975 and even those were the responsibility of the IWC,

In 1993, Norway shocked the International Whaling Commission with its unilateral resumption of commercial whaling.

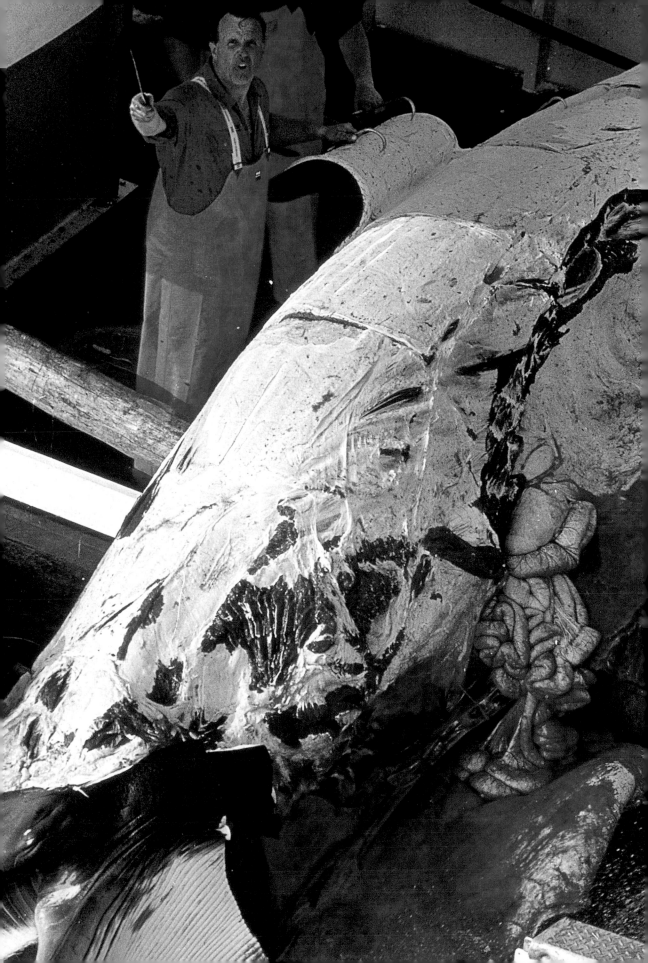

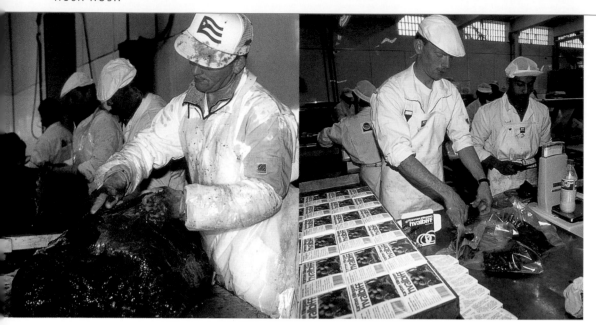

Whaling and the processing industry that relies on it are long-standing traditions in Norway. Here, whales are skinned and fleshed at local plants.

The Norwegian authorities claim that whale meat is sold only for domestic consumption in Norway, but evidence reveals that the law is being flouted by exporters.

not the Norwegians. In other words, Norway was seen by many in the international body to be operating on the principle of hunting whales while the going was good and letting the future take care of itself.

For a time, the Norwegians were prepared to abide by the IWC ban and even supported the findings of the IWC Scientific Committee when in 1991 it established that Minke stocks in the northeast Atlantic had suffered a major decline. So on what grounds did Norway justify its apparent flouting of the efforts just a year later? The change of policy relied on an estimate presented to the IWC, which on closer inspection turned out to be based on highly questionable data. In 1992, Norway came up with figures indicating that the Minke population of the northeast Atlantic stood at just under a healthy 87,000. At first, the commission accepted these figures, although in hindsight the IWC was guilty of a naivety, since the data the estimate was made from had been prepared was not submitted at the time.

Revealing the truth

When non-Norwegian scientists were eventually able to scrutinize the calculations, it became apparent that serious doubts must be cast over their accuracy. By 1995, a working group analyzing the data revealed that the computer model used by the Norwegians contained major errors. By that time it was too late. To the immense concern of the majority of IWC members, Norway had resumed the hunting of Minke. Seventeen IWC member states initially signed a joint statement declaring that the action might "seriously undermine the only international organization with authority to manage whaling." Twenty-one nations called on Norway to "halt all whaling activities under its jurisdiction" in 1995.

The Norwegian government ignored the plea. It responded that whaling was a long-standing tradition doing no harm to the whale population. Arguably, however, the decision was made more for domestic political gain than out of serious consideration for the welfare of whales. Local fishermen had frequently complained that whales and seals were to blame for declining fish catches. Norway was eager to promote the image of a cozy cottage industry, whereas in reality the modern Norwegian whaling boats are more like factory ships, weighing up to 300 tons and questing as far as the Russian whaling grounds.

Abuse of regulations by the Norwegian whalers has always been a matter of contention. It may have already been widespread, if largely undetected, before 1992, since the limited number of whaling inspectors who were employed during the 1980s reported frequent abuses of quotas that the vessel owners largely got away with. The situation has scarcely improved. In 1994, the owner of the whaling ship *Dag Senior* was fined and had his catch confiscated for breaking quota limits after having been reported by another vessel, but actual prosecutions of this kind are rare.

Furthermore, the claims of "home consumption only" have been proved nonsense. In October 1993, a shipment purporting to consist of frozen prawns bound for Japan, having been cleared for export by customs staff at Oslo airport, was found to contain whale meat when one of the boxes accidentally split. On January 16, 2001, under industry pressure, the Norwegian government reversed the ban on export of whale products, effectively ending an international agreement that had existed since 1986.

The question has to be asked whether the Norwegian scientists who produced the data knew that the computer model contained serious errors, which the IWC found patently obvious when allowed access to the details? The Norwegian newspaper *Aftenpost* probably came closer to the bones of the matter on June 7, 2000. It reported that coastal voters were being urged to sign a petition to the minister of fisheries demanding the legal export of whale products. Those pressing for change in the law were the same activists who had clamored for a resumption of whaling in 1992. The issue, the campaigners stated, "does not only concern profitable jobs for whalers but for all fisheries related fields. As long as we have a self-imposed ban, whaling will not be profitable."

The 1992 computer model, on which Norwegian scientists argued that the Minke whale stocks are at a healthy level, is seriously flawed. We simply do not know how many Minke whales such as this one currently exist in the northeast Atlantic.

REFERENCES

IMMP responds to Norwegian whaling claims. *Earth Island*. Undated.
www.earthisland.org/immp/whlclaim.html

Moy, R. (Engl. transl. Krag, A.) This is a plead [sic] to everyone who is concerned with Norwegian whaling. *Aftenpost*. June 7, 2000.
www.dyrebeskyttelsen.no/english/whaling.htm

Norwegian Minke whaling: coastal livelihood and natural resource management. Odin. *Ministry of Foreign Affairs*. Undated.
http://odin.dep.no/odin/engelsk/norway/environment/ 032001-990108.

Norwegian whaling-neither small scale nor traditional. *Greenpeace*. Undated.
www.greenpeace.org/~comms/cbio/norweg.html

5.6 Atomflot: A Catastrophe-in-Waiting

Wrecks at the entrance to Murmansk harbour on the Kola peninsula, Russia. Note how close these wrecks are to the apartment buildings on land.

In 1996, the Russian government of Vladimir Putin put a former Soviet submarine captain on trial for treason after he "blew the whistle" on the dangerous state of the obsolete Russian nuclear fleet.

The name of Alexander Nikitin may not be familiar to many people in the West, yet his story is horrific in its implications. Nikitin is a former Soviet submarine captain and naval specialist in nuclear safety who left the navy in 1992 and was put on trial by the Russian Putin government in 1996 charged with "acts of treason." His alleged crime was to have blown the whistle on the growing dangers aboard obsolete Russian nuclear submarines, part of the Northern Fleet that now lies aging and derelict around Murmansk and the Kola Peninsula, an area accurately described as the world's biggest nuclear trash can and euphemistically referred to as Atomflot. According to a BBC report, the coastline is littered with rusting nuclear submarines, a fifth of the world's entire stockpile of nuclear reactors and a concentration of nuclear fuel. Radio Murmansk puts out daily bulletins not only for the weather forecast but also for radiation levels.

Under the sponsorship of Norwegian environmental group Bellona, Nikitin documented the risks posed by Atomflot. In September 2000, he was acquitted by the Presidium of the Russian Supreme Court, but the ordeal he underwent reveals the reluctance of the Russian government to come clean about its very dirty nuclear arsenal and the risks that this currently poses to the environment. His report has been banned in Russia and it has been revealed that some of the background material was confiscated by Russian security police during a raid on Bellona's Murmansk office, hindering the completion of the report.

The horrors revealed

The risks described in the report are many and various. The first storage facility for spent nuclear fuel on the shores of Andreeva Bay, situated on the west side of Zapadnaya Litsa fjord, is known only as "Building 5."

It was erected in 1962 and enlarged in 1973. Inside the concrete shell, the storage area consisted of two 200-foot (60-meter) long water-filled pools lined with steel plates into which fuel rods from all the Northern Fleet shipyards were packed in containers. In theory, the water level should have been monitored from an adjacent building, but because of technical problems, the inspections had to be carried out manually. In the spring of 1982, it was noticed that Tank 2 within Building 5 had begun to leak. By the autumn of that year, despite efforts to resolve the problem, the tank lining had deteriorated to such an extent that up to 30 tons of radioactive water was escaping into the sea each day and the level inside the tank was dropping so severely that there was a risk of the storage containers becoming exposed. Then the other pool began to leak, losing up to 10 tons of water a day. It took until 1989 to remove all the fuel rods from Building 5, and the radiation levels around it are still dangerously high. It has been calculated that 3,925 cubic yards (3,000 cubic meters) of radioactive water escaped from both pools into the fjord. Yet, in spite of the severity of the situation, the first information did not reach the outside world until 1993. Even at that juncture the Russian government still delayed an official confirmation of what had taken place.

By 1996, thousands of radioactive fuel rods had been packed into three "dry storage" concrete tanks situated below Building 5. These are now completely filled and are said to be in poor condition. More rods have been stored in containers in the open, also deteriorating.

One of the most severe risks, aside from Building 5, is probably that of a 60-year-old ship, the *Lepse*, which lies off Murmansk. The vessel is used to store spent nuclear fuel and, according to a BBC investigation in August 2000, at that time contained 642 bundles of fuel rods, two-thirds of which were damaged and still hot. It has been alleged that the Russians have tried to cram so many fuel rods into the vessel by physically hammering them down that the superstructure has buckled.

A time bomb

Nikitin's disclosures have clearly embarrassed the Russian government. The report stresses that without international cooperation and financing, a "slow-motion" Chernobyl situation is waiting to happen. Without urgently needed safety measures, major accidents and the release of

 The transferring of spent fuel rods from a nuclear icebreaker to the storage ship Lotta in Murmansk, Russia, part of a burgeoning stockpile of dangerous radioactive rubbish.

A worker measures radiation levels using a geiger counter.

radioactive material will be unavoidable. Yet far from Nikitin having transgressed the laws of his country, it is actually the maintenance of secrecy over information of this nature that is in contravention of Russian law. Article 7 of the legislation covering State Secrets (1993) establishes that "Information on the condition of the environment is not subject to classification." Article 10 of the legislation on Information and Protection of Information (1995) states: "It is prohibited to ascribe the following to materials with limited access... documents which contain information on extraordinary situations, environmental information and other information necessary to ensure the safe functioning of residential areas and industrial sites."

As Nikitin told the BBC: "I have decided to talk about the submarines because it is a real danger, it is a real problem...I wanted to draw attention so that it is dealt with." Atomflot is not the only real problem over which Russia has preferred to maintain secrecy. According to the CIA, at the end of 1997 Russia possessed 40,000 nuclear warheads, 12,000 tons of enriched uranium and 200 tons of plutonium kept at 50 sites scattered throughout Russia. In the summer of 2002, Vladimir Putin agreed with the United States to further reductions in the stockpile of nuclear weapons, but without another Alexander Nikitin to blow the whistle, we can only guess at the state in which the rest of the arsenal is kept.

The coastline of the Kola Peninsula is said to be littered with rusting nuclear submarines like this one.

REFERENCES

Bellona. The Russian Northern Fleet. Report 2. Nuclear Submarine Accidents. *TEIA Web Site*. Undated.
www.spb.org.ru/bellona/ehome/russia/nfl8.htm

Persecuted environmentalist triumphs despite Putin crackdown. *Amnesty International*. September 13, 2000.
www.amnesty-usa.org/justearth/updates/v_russia2.html

Robbins, James. Russian nuclear dustbin threats. *BBC News Online*. August 14, 2000.
http://news.bbc.co.uk/hi/english/world/europe/newsid_607000/607175.stm

Russian fleet destroys ballistic missiles. *BBC News Online*. December 4, 1997.
http://news6.thdo.bbc.co.uk/hi/english/world/monitoring/newsid_36000/36961.stm

The Russian Northern Fleet. Preface. *Digital Freedom Network*. Updated June 10, 1999.
www.dfn.org/voices/russia/nflo-1.htm

The Russian Northern Fleet. Report 2. Radioactive waste at the naval bases. *Bellona*. Undated.
www.bellona.no/en/index.html

5.7 Space Debris

The effects of collision between space craft and even small particles of debris is very severe, as this photograph reveals.

Between June 1, 1994, and May 31, 1995, the most recent period for which figures are readily available, the Russian Space Agency alone blasted 19 proton-type rockets into the Earth's orbit, which weighed a combined total of more than 220,000 pounds (100,000 kilograms). Much of this material then became redundant and now represents just a small part of the overall statistics relating to space debris. Current estimates on the number of bits and pieces of man-made "space junk" orbiting the Earth vary somewhat, but in March 2001 the European Space Agency (ESA) came up with a figure of approximately 8,500 trackable objects larger than four inches (10 centimeters) in diameter. In May of that year, the BBC put the total at nearer to 10,000. Other ESA statistics point to 50 percent of the larger objects consisting of decommissioned satellites, spent upper stages of launch vehicles and other "mission-related objects." These figures, however, do not include literally tens of millions of smaller pieces. Most of these originate from the more than 130 "disintegrations" that have been recorded since 1961, all but a handful resulting from explosions of spacecraft and upper stages of rockets.

A little does a lot of damage

Small orbiting debris, even as little as a fraction of an inch in width, may appear to be of little concern. But when traveling through space at a speed of 22,320 miles (36,000 kilometers) per hour, a little nugget of aluminum 0.39 inches (one centimeter) across will impact on a satellite or a space shuttle carrying a human cargo with a force comparable to a steel safe weighing more than 400 pounds (180 kilograms) impacting at 62 miles (100 kilometers) an hour. Even tiny flecks of paint moving through space at that kind of velocity frequently damage the windows of space shuttles. Furthermore, if a larger fragment of debris, say four inches (10 centimeters) across, hits a properly shielded spacecraft, the fragment will disintegrate into more than a million pieces, creating a cloud of tiny particles hazardous to any other craft passing in the vicinity.

The space shuttle *Endeavour* lifts off into space. But some of its costly cargo is destined to become tomorrow's space junk.

According to recent scientific reports, NASA and the agencies of other nations are figuratively, and perhaps literally, shooting themselves in the space boot. The situation has now become so bad that, even if we added no more debris to the existing volume, we are jeopardizing future near-space access because it is liable to become too dangerous. Pieces of debris will impact with other detritus in orbit around the Earth and this will progressively form a dense cloud containing millions upon millions of fragments, each potentially lethal to a space ship or astronaut of the future. By and large, the dangers thus posed affect only the safety of astronauts because the chances of large lumps of this debris coming to Earth and hitting us on the head are very small. On the other hand, they pale into insignificance by comparison with the risks from man-made objects in space about which we are told very little. The concern is not so much about the volume but the nature of the material.

In November 1996, the Russian CIS space agency launched a deep-space probe destined for Mars. Its final-stage booster failed to ignite and the craft plunged back to Earth, allegedly impacting somewhere in the Pacific Ocean. The probe is said to have contained between 200 and 300 grams of plutonium, a highly radioactive substance used as the power source for its interplanetary journey. In October 1997, NASA launched a deep-space probe called Cassini aboard a Titan IV rocket, a vehicle with a 10 percent failure rate. Its destination is Saturn, but Cassini first needed to increase its speed. In order to achieve this, a "slingshot" effect needed to be created, so it was placed in an elliptical orbit first around Venus, then around Earth. On June 24, 1999, having visited Venus, the probe headed back toward Earth at approximately 43,000 (71,600 kilometers) miles per hour, passing us at a distance of 725 nautical miles. According to NASA's original plan, subsequently revised, Cassini should have passed a mere 312 miles above the Earth's surface. As has been pointed out by critics, one minor miscalculation or malfunction and a chance impact with one of the pieces of existing space debris could have altered the trajectory and sent the spacecraft plunging to Earth, a danger that NASA

attempted to minimize. What was not revealed to the public is that Cassini is carrying, again as a power pack, a little less than 73 pounds (33 kilogram) of plutonium dioxide. Plutonium is a lethal substance with a half-life of 250,000 years. Medically, it is well known that when even a tiny particle of plutonium is inhaled, the localized radiation can cause lung cancer. The plutonium in Cassini was more than NASA has ever put into space. Had the launch or the subsequent fly-by resulted in disaster, with the power pack being pulverized in an impact, large quantities of small-particle-size plutonium would have been dispersed into the atmosphere.

A computer illustration of an explosion on space shuttle *Atlantis* due to a collision with space junk. Such an impact would almost certainly be fatal for the crew.

A regulation-free zone

In space, there is no equivalent to the U.S. Environmental Protection Agency, and if, as it seems from the evidence, we are indeed creating a gigantic garbage dump out there, very little exposure has been given to the problem. In terms of deep-space probes, not only do we risk our own health and safety on Earth, but also any kind of disaster at the "other end" of the journey could result in our inadvertently polluting a sister planet for a very long period of time. The biggest culprits, the military authorities in countries like the U.S. and Russia, are responsible for generating vast quantities of debris in space, yet they are answerable virtually to nobody, and the public is given little information about what takes place and the possible risks. Are governments with space exploration capabilities contributing to an environmental disaster of the future?

REFERENCES

Anselmo, L. Launch rate of rocket bodies (>900kg) (1 June 1994-31 May 1995). *CNUCE Spaceflight Dynamics Group. SMOS Reports.* Undated.
http://apollo.cnuce.cnr.it/REPORTS/Rocket_Bodies.html

Hoffman, R. D. A plutonium-powered Russian space probe falls to earth. *The Animated Software Company.* 1996.
www.animatedsoftware.com/spacedeb/mars9611.htm

Hoffman, R. D. The problem of space debris. *High Tech Today.* Undated.
www.animatedsoftware.com/spacedeb/spacedeb.htm

Hoffman, R. D. *Stop Cassini Home Page.* The Animated Software Company. Last modified April 2002.
www.animatedsoftware.com/cassini/index.htm

Klinkrad, H. H. Space debris activities at ESOC. *European Space Agency.* Last updated January 23, 1996.
www.esoc.esa.de/external/mso/debris.html

The Third European Conference on Space Debris. *European Space Agency.* March 19-21, 2001.
www.esoc.esa.de/pr/conferences/conference.001.php3

What are the risks posed by orbital debris? *The Aerospace Corporation.* Last modified February 23, 2001.
www.aero.org/cords/debrisks.html

Whitehouse, D. Space debris warning. *BBC News Online.* May 25, 2001.
http://news.bbc.co.uk/hi/english/sci/tech/newsid_1351000/1351233.stm

6 Saving Face

6 Saving Face

Saving face can be an important factor in cases of scientific secrecy. It is human instinct that if we have made a mistake, whether accidentally or by design, our first reaction may be a defensive one to avoid being found out.

A radiologist examines X-rays mounted on lightboxes. It was not until 1970 that the medical establishment admitted the dangers of X-rays and implemented compulsory safety procedures.

We attempt to cover our tracks in a variety of ways. We can deny involvement while attempting to conceal the evidence that we are at fault, or we can pass the buck by blaming someone else for our mistakes. As children we must all, at one time or another, have protested that "it wasn't me, it was him!" It is a ruse we are capable of carrying into adulthood. Professional pride, and the worry that an individual or a corporation can be found culpable in terms of negligence or worse, lends itself readily to the secrecy game, except it isn't a game anymore. The coverup can quickly reach unacceptable and dangerous levels. The consequences of trying to escape detection can be infinitely more serious than merely landing a school friend in trouble while wriggling out of punishment.

The fervent desire of a corporation management team, an individual scientist or a technician to escape the consequences of laxity or more deliberate wrongdoing may encourage an assortment of measures to seek self-protection. Not only do these damage-limitation exercises frequently result in more severe problems for the victims of error, but they can also adversely affect the careers of fellow workers. The guilty parties pressure the innocent into keeping quiet, and one of the means is to threaten the future security of the "whistle-blower." Employees may easily become victims of failures and poor judgement by someone higher up the corporate ladder. By speaking out, they may risk losing their jobs, their futures, or worse.

In the U.K., one of the most recent examples where whistle-blowing resulted in dismissal came in the aftermath of a series of train crashes. A senior incident officer who had worked for the railway authority spoke out on standards of track repair in the location of a disaster scene and published photographs that appeared to show accountable deterioration of wooden track ties. He asserted in a television interview that he was to be barred from attendance at future accident scenes associated with the railway, despite his extensive experience in such matters and an unblemished record. Here was information that might have nothing to do with a recent train crash but had the potential to show up the railway authority as being incompetent, so diminishing its credibility in the eyes of the public. Such reactions illustrate another trait that we first demonstrate in childhood, the principal that no one likes a sneak. Yet in the adult world, the failure to expose our colleagues can bear more sinister connotations and consequences.

The complexity of some coverups can make it difficult to define the true nature of the beast. In these cases of commercial and medical secrecy, the perpetrators consciously held information back from the critical public gaze for fear of losing their reputations, their future business, and, as individuals involved in the cases, their careers. So much of today's world of commerce relies on the reputations of the companies and individuals involved, particularly in the fields of medicine and transport, that the mere idea of bad publicity is enough to warrant a massive coverup. But at what cost are these decisions made? The victims and their families can take no solace from the reasons the companies and organizations involved give.

Aerial photography at the site of the Air France Concorde crash near Paris reveals just how close the supersonic aircraft came to a large hotel. Over 100 people died in the accident.

6.1 The Ferry Disasters

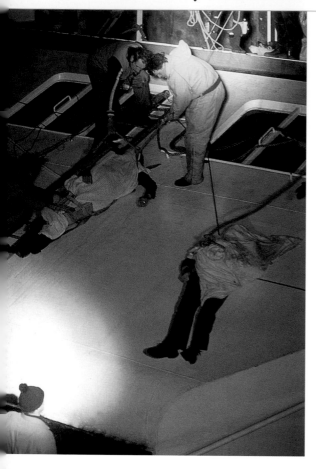

On the morning of March 6, 1987, the roll-on, roll-off ferry (ro-ro) *Herald of Free Enterprise*, bound for Dover carrying a crew of 80, 459 passengers, 81 cars, 47 trucks and three buses capsized shortly after leaving the Belgian port of Zeebrugge. Run by Townsend Thoresen, which had only just become a part of P&O Ferries, the ship was modern, designed for use on the busiest cross-channel routes, and certified to carry up to 1,400 passengers and crew. The sea was calm with only a light easterly breeze blowing. Yet within about 90 seconds, the ship was lying on its side in shallow water in the approaches to the harbor, resulting in the deaths of at least 188 passengers and crew from drowning and hypothermia, most of them trapped inside the hull.

Speed over safety

Some of the biggest ferries, including the *Herald of Free Enterprise*, have been built with upper and lower vehicle decks, which are accommodated at Dover, England, and Calais, France, with double-deck quayside ramps. Zeebrugge, however, possessed only a single ramp, so to access the upper deck, the bows of the ship had to be lowered by pumping ballast water into bow tanks. The emphasis on fast journey times had also led to the *Herald of Free Enterprise* being designed to accelerate rapidly to a service speed of 22 knots. On the morning of March 6, before all the water had been pumped from the ballast tanks, the ferry backed out stern first, turned and raced for the open sea while still some three feet down at the bow. The horizontally hinged bow doors, invisible from the bridge, were also open. At 15 knots, water began to break over the main vehicle decks, entering the vessel at a rate of 200 tons per minute.

Rescuers recovering the bodies of some of the passengers who died in the Zeebrugge ferry accident.

A bitter sweet irony

In a letter published in April 1987, introducing the 1986 annual report of P&O, the company chairman Jeffrey Sterling made the assertion that "You may be assured that the safety of our ships and those who man them and travel in them is our overriding priority." The report of the formal investigation, however, told a different story; that "from top to bottom the body corporate [of Townsend Thoresen] was infected with the disease of sloppiness" and that the standards of management reveal "a staggering complacency." P&O earned a damning indictment. The defense counsel for the company had claimed, when asked who was responsible, "Well, in truth, nobody, though there ought to have been."

The Herald of Free Enterprise after it capsized soon after leaving the port of Zeebrugge.

The investigators placed the blame fairly and squarely on the company. They concluded that the directors had no proper comprehension of their duties and should accept a heavy responsibility for their lamentable lack of direction to those working lower down the chain of command. As the investigation proceeded, it became apparent that the company's Marine Department had virtually ignored advice, complaints and suggestions regarding safety given by the ships' masters in at least four instances. Meetings between masters and management had been intermittent, in one instance involving a lapse of four and a half years.

Yet officials' explanations about the cause of the *Herald of Free Enterprise* sinking appear to fall well short of telling the entire story. Modern ro-ros are essentially built as pontoons with a superstructure and doors at either end. Vehicles roll on at one end and roll off at the other, hence the name. Prior to 1987, such vessels were built without subdividing bulkheads on the main vehicle decks. Construction costs were thus minimized and the flow of vehicles was not impeded. Nor, unfortunately, was the potential flow of water. The absence of such bulkheads facilitated the free movement of seawater from side to side and from one end of the vessel to the other. Despite concerns, this factor appears to have gone largely unheeded, and in September 1994 another roll-on, roll-off ferry disaster was destined to take place. The Scandinavian car ferry *Estonia* sank in not dissimilar circumstances. Its bow "visor" became detached, the front-loading ramp opened and seawater flooded the vehicle deck causing the ship to become critically unstable.

These accidents were not the first to beset the car ferries. Less well publicized than the disasters of the *Herald of Free Enterprise* and the *Estonia*, is that of the *European Gateway*, a British ro-ro carrying only crew and freight, which capsized and sank in 1982 after collision with another vessel. The cause of the sinking, like that of the later ferries, was attributed to the flooding of the *European Gateway*'s car deck.

The ferry industry has been accused of drawing attention away from the design shortcomings of its ships. Shortly after the Zeebrugge disaster, the records of the U.K. House of Commons reveal that then Prime Minister Margaret Thatcher stated her understanding "that it was a fundamental design of these vessels that was the problem, and that something would have to be looked at very quickly to reassure the public." Yet within two days of her statement, the U.K. government was taking quite a different line, allegedly having been "got at" by the ferry industry. The Secretary of State for Transport asserted, "The loss of the *Herald* [*of Free Enterprise*] was not due to design problems. It was entirely due to operational error."

In the official Hansard minutes of a parliamentary debate on the matter of ferry safety after the loss of the *Estonia*, dated February 21, 1996, it is pointed out that human error causes 73 percent of all marine accidents. In the case of the British ro-ro ferry disaster of 1987, however, it was alleged by the backbencher naming the debate, Paul Flynn, that

Jeffrey Sterling CBS, Chairman of P&O in 1987. The formal investigation into the *Herald of Free Enterprise* disaster charged the company with sloppiness and staggering complacency in its management standards.

Wreckage on the *Herald of Free Enterprise*. Were dangerous design shortcomings responsible for the disaster?

the prime minister was originally right to identify the problem as a design fault and that the British government then decided to change their reaction under pressure from the ferry industry. Flynn refers to "an incredible, sorry story of government complacency toward compliance of the maritime industry. This is a story of an industry that has, over many years—with the government's help—pursued minimum safety standards in search of maximum profits."

Flynn's Commons statement of 1996 revealed that 13 years after the first sinking of a British ro-ro, the Royal Academy of Engineering had said that the solution to the fundamental design fault in such vessels was still not being implemented for ferries built before 1990 and only marginally for some built since that time. What is of outstanding worry in terms of secrecy is not so much the lackadaisical attitude of the maritime companies toward safety standards, though this in itself is reprehensible enough, but the alleged smokescreen put up by the British government to draw attention away from dangerous design shortcomings in the vessels in order, it has been claimed, to appease pressure from the ferry operators.

Families of some of the victims of the *Herald of Free Enterprise* accident.

REFERENCES

Bjorkman, A. Lies and truth about the M/V Estonia accident. Chapter 5. January 1998. *The Heiwa Co.* Updated to the Net, May 2000. **www.heiwaco.tripod.com/news.htm.**

Boyd, C. Herald of Free Enterprise car ferry disaster. *Case Studies in Corporate Social Policy in Post, Frederick, Lawrence & Weber. Business and Society (New York).* 1996.
http://business.unisa.edu.au/corpresp/case_studies/study3.htm

Hansard. Extracts from the House of Commons minutes for February 21, 1996: Columns 281-284.
www.parliament.the -stationery-office.co.uk/pa/cm199596/cmhansr/60221-01.htm

6.2 X-ray Secrecy

Most people in Britain over the age of 50 will remember the experience of going to the shoe shop and standing with their feet in a small recess at the base of an impressive looking machine known as a fluoroscope. They peered into a viewing screen and their foot bones miraculously appeared in an eerie greenish light. They were witnessing an everyday application of the properties of X-rays. The fluoroscope had become the accepted way to ascertain if the shoes of a growing youngster were fitting properly, and it has been estimated that at one stage in the 1950s, as many as 10,000 such machines were in use. Yet X-ray machines represented a threat to health that no one realized at first, and even when the danger was exposed, many radiologists continued to insist on X-ray treatment being safe. It is arguable that self-interest on the part of practitioners not wanting to be proven wrong fostered a culture of covering up that could otherwise have saved many people from developing life-threatening diseases.

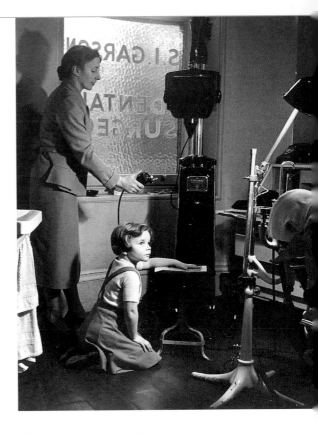

An amazing invention

In 1895, Wilhelm Roentgen made the extraordinary discovery that he could generate an invisible electromagnetic ray, a kind of radioactivity that would penetrate matter and permit inspection of the internal structure of the human body. These rays became hailed as a miracle weapon in the fight against disease and injury. With Roentgen rays, or X-rays as they became popularly labeled, it was possible to see a hairline fracture in a bone, detect signs of lung damage from tuberculosis and, of course, find out if little Johnny needed a new pair of shoes. Medical practitioners rushed to obtain X-ray machines and used them at the least excuse to determine the nature of all kinds of injuries, no matter how trivial. Patients, too, tended to expect X-rays to be taken, much as they expect to walk away from the doctor's office these days with a prescription in their hands.

A girl having her hand X-rayed in 1952 before the dangers of radiation were fully understood. Safety precautions were few and far between.

Soon, medical experts were looking for new opportunities to use the revolutionary technology. In the years immediately before World War II, it became clear that repeated doses of X-rays affected certain body tissues and organs; among the effects was shrinkage of the thymus gland in infants. Medical research had already identified the thymus to be closely involved with development of the immune system during the first months of life. In a newborn baby, the gland is large, but it then shrinks to almost nothing after a year, by which stage the child has become resistant to passing infections. In the late 1940s, the untested rumor spread in the medical profession that subjecting infants to radiation therapy could

Professor Wilhelm Conrad Roentgen (1845–1923), who discovered the X-ray.

Marie Curie, who won the Nobel Prize for Physics in 1903 and the Nobel Prize for Chemistry in 1911. Curie was the first person to discover the radioactive elements polonium and radium. She died of leukemia, which is now linked to her work with radium.

boost their immune systems. Thus, babies suffering from nothing more life threatening than the common cold were treated with doses of X-rays. Later, the same technique came into vogue to shrink adenoids and tonsils, and even to clear up acne.

Doctors dismiss the research

In 1950, however, two doctors, B. J. Duffy and P. J. Fitzgerald, analyzed records of 28 children who had thyroid cancer, two of whom had already died. It was discovered that nine of these patients had been subjected to X-rays when less than a year old in order to shrink their thymus gland. When the report was published in the medical journal *Cancer*, it was heavily criticized by those in the profession who supported X-ray therapy and attempts were made to discredit the two doctors on the grounds that their sample was far too small to be meaningful.

Nevertheless, doubts remained and in 1952 a second paper was published by two Americans, Goldberg and Chalkoff, showing that radioactivity in their experiments emitted from radioactive iodine encouraged the development of thyroid cancers in rats. They drew the conclusion that the earlier conclusions of Duffy and Fitzgerald were correct, but once more, others in the medical establishment poured scorn on their work, asserting that experiments on rats did not prove anything about the effects of radioactive substances on human beings.

They finally give in to the truth

It took until 1970 for the critics of X-ray therapy to persuade the diehards about the lethal potential of X-ray machines. A study published in the *New England Journal of Medicine* revealed that 15 percent of thyroid cancer patients taking part in a large population study had as small children received X-rays designed to shrink various glands. By the end of the 1960s, the number of people placed at risk had grown out of all proportion. A single study in Chicago established that the health of as many as 70,000 in the area of the city might have been affected. It wasn't until 1973 that X-ray machines were withdrawn from general use and X-ray dosages were strictly governed so as to minimize risk to the recipients.

Recent evidence indicates that medical X-rays are a major cause of human cancers and of coronary heart disease. The Center for Devices and Radiological Health in the U.S. Food and Drug Administration has also stated that one-third of all diagnostic X-ray examinations still routinely carried out are unnecessary and do not contribute to the medical benefit of the patient. A large number of the remaining two-thirds deliver a higher dose of radiation to the patient than is necessary for obtaining complete diagnostic information. In May 1996, the Australian Broadcasting Corporation screened an investigation claiming that in Britain, doctors were still underplaying the risk of X-ray treatment and putting patients' lives at risk.

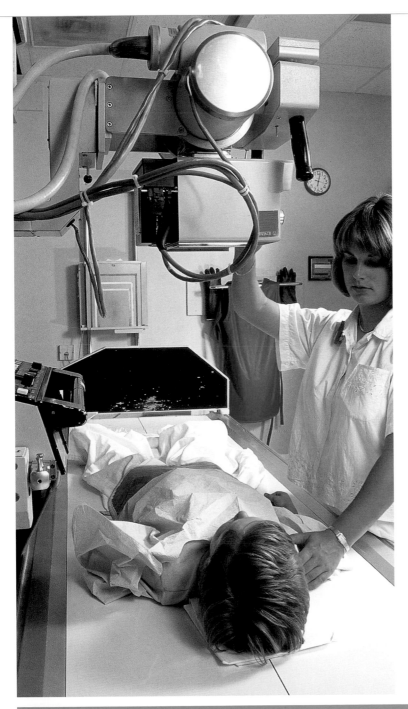

Modern hospital radiography has evolved so that doses of X-rays are now much lower than in the past. The radiologist will go behind a protective screen before administering the dose to her patient.

REFERENCES

Cause of thyroid carcinoma and nodules. Chapter 18. *Thyroid Manager.* May 14, 2002.
www.thyroidmanager.org/chapter18/18-cause.htm

Footnotes. Chapter 7. *ACHRE Report.* Undated.
http://tis-nt.eh.doe.gov/ohre/roadmap/achre/chap7_fn.html

Risk of harm and nontherapeutic research with children. Chapter 7. *ACHRE Report.* Undated.
http://tis-nt.eh.doe.gov/ohre/roadmap/achre/chap7_3.html

The X Ray files. Four Corners. *Australian Broadcasting Corporation.* May 20, 1996.
www.abc.net.au/4corners/stories/s72923.htm

6.3 Concorde

July 25, 2000: Flames come out of the Air France Concorde seconds before it crashed in Gonesse, near Paris' Roissy airport.

In the summer of 2000, through TV, the world witnessed the dreadful sight of an Air France Concorde taking off from Charles de Gaulle Airport, Paris, trailing a stream of ignited fuel and exploding in a fireball. More than 100 people were killed. To what extent was the public ignorant of the potential danger of flying by Concorde? On the face of it, there was little to worry about. Concorde overtly boasted an untarnished safety record, and the fleet of supersonic airliners, with their striking appearance of slim fuselage, drooping nose and razor-thin wings, had flown millions of miles for more than 30 years, often faster than the speed of a rifle bullet. A former director of NASA once described the complexity and success of the technology as comparable to placing men on the surface of the moon.

Design shortcomings

Yet the same sophisticated science included a hidden weakness that was ultimately to prove disastrous. To understand its nature requires a brief dip into the technology of Concorde's abilities. When the aircraft travels at such enormous speeds, considerable friction is created as air moves over the surface of the wings and fuselage. Unless corrected, this friction would cause the aircraft to heat up to such temperatures that its metal skin would be at risk of melting. To counter the problem, an efficient system of cooling had to be devised. The solution was to use the aviation fuel, enormous

France's Air Accident Investigation Bureau released a preliminary report including a picture of a metal strip, probably from another plane, which was found at the runway at Paris Roissy airport following the horrific crash of the Concorde. Since the start of the investigation, experts have speculated that this 43cm strip caused the plane's tire to explode.

The Air France Concorde crash site. What little remains of the aircraft after the crash is extinguished by firefighters.

volumes of which are contained in 10 tanks dispersed through the delta wing with a further three located in the fuselage. During flights, the fuel supplies the four Olympus engines but, different from conventional airliners, it is also circulated through pipes to help cool the aircraft. In addition, it is pumped backward and forward during different phases of the flight to adjust the center of gravity, a factor that changes according to whether the plane is traveling at supersonic or subsonic speeds.

It was not widely appreciated, however, that the technology of pushing combustible fuel around the aircraft left it vulnerable to any accidental impact that might rupture the skin and penetrate the fuel circulation system. Nor was it publicized that Concorde had a history of tire problems because of the heavy loads it carried and the high landing and takeoff speeds. On November 9, 1981, the National Transportation Safety Board (NTSB) in Washington, D.C., wrote to the French authorities warning them that, on four occasions during a 20-month period between July 1979 and February 1981, Air France Concordes operating from Dulles International and Kennedy International airports had been involved in potentially catastrophic accidents resulting from blown tires during takeoff. A double tire burst was recorded on July 14, 1979, at Dulles International airport when debris punctured three of the fuel tanks in the port wing and cut electrical cables and hydraulics, causing a fuel leak; blown tire incidents occurred at Dulles on July 21, 1979 and again in 1981. There was also an incident of tire bursts at Kennedy airport in October 1979. The NTSB advised that "the repetitive nature of these incidents and, in particular, crew response in the more recent incidents is of serious concern." On two occasions, the crews raised the landing gear anyway despite the blown tires. One flew on to Paris while another was forced to land at New York because of engine damage.

Disaster strikes
On July 25, 2000, a chain of events was initiated when a subsonic commercial aircraft, using the same runway that Concorde would take at Charles de Gaulle, accidentally deposited a thin 25-centimeter strip of

As seen in the top illustration, when the tire of the Air France Concorde exploded, fragments shot up into the left wing, rupturing the fuel tanks, which almost immediately set on fire. The bottom illustration shows the location of the engines and fuel tanks, through which the fuel was constantly moving to keep the aircraft cool.

Engine 4

Engine 3

Engine 2

Engine 1

Engine feed banks
Main transfer banks
Trim transfer banks

Jean-Cyril Spineta, chairman of Air France, faces the press after the Paris Concorde crash. Yet questions remain unanswered over why the company did not heed warning issued by the U.S. National Transportation Safety Board until after disaster struck.

metal onto the concrete. When the Concorde flight approached its takeoff speed of 200 miles (320 kilometers) per hour, metal fragments from the strip cut into and burst one of the tires, hurling chunks of rubber weighing up to nine pounds (four kilograms) apiece into the wings. They punctured the fuel tanks, causing a massive leakage onto the hot engines, and the port wing caught fire. Within minutes, all 109 passengers, the crew and four victims on the ground were dead.

In consequence, the Concorde fleet was grounded until November 2001, during which time the safety improvements that were demanded cost British Airways some £17 million ($27 million), in addition to its loss of income from suspending the service. The fuel tanks are now reinforced with Kevlar armor protection panels, as is the undercarriage wiring, and the tire technology has been radically redesigned by Michelin, not the original manufacturer of Concorde's tires, to withstand high-speed impact without bursting. The questions remain over why, on the face of it, the British and French airline authorities did not heed the warning issued by the NTSB until after a major disaster had occurred, and why the NTSB did not release details of the risks being taken by Concorde passengers until July 2000.

REFERENCES

US points to Concorde tire problems. *BBC News Online.* July 28, 2000.
http://news.bbc.co.uk/hi/english/world/europe/newsid_856000/856173.stm

US warned on tire risks. *BBC News Online.* July 28, 2000.
http://news.bbc.co.uk/hi/english/world/europe/newsid_856000/856423.stm

The Human Cost

Violating human rights in the name of science

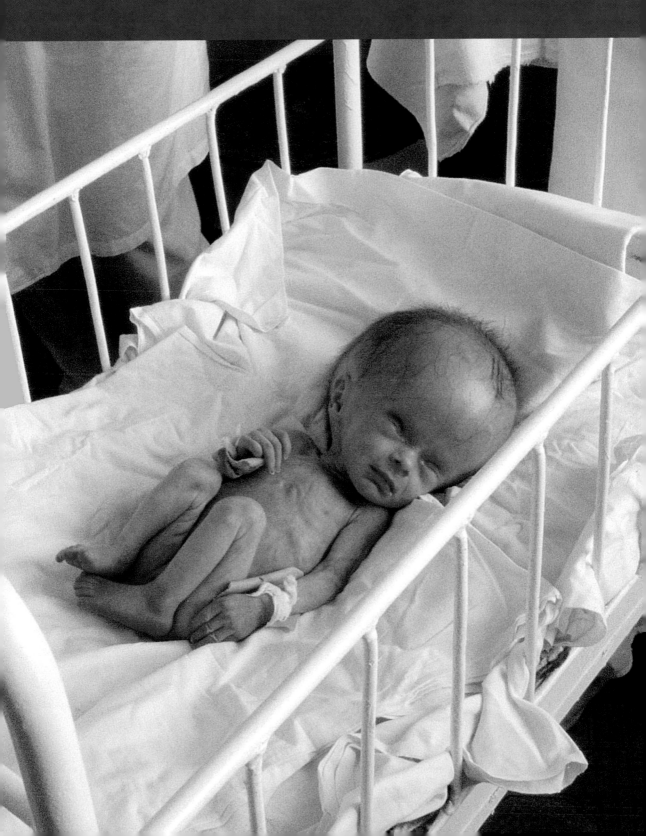

7 The Human Cost

Few people would argue against the opinion that the most intolerable aspects of scientific secrecy are those that infringe on and violate human rights. Science can be used for good or evil intent and there have been far too many cases in the last hundred years when scientific knowledge has been exploited for the wrong reasons, causing human misery. Anyone who has seen television reports of Iraqi Kurds who have been exposed to chemical weapons or the agony of the citizens of Bhopal when corporate negligence resulted in the release of a cloud of poisonous gas over the town, will have little sympathy for the perpetrators.

Hermann Goering at the Nuremberg trials, which were held between 1945 and 1949 to try leading figures in the Nazi party for war crimes during World War II. Doctors and scientists were among those prosecuted for forced experimentation on human beings in labor and concentration camps. The trials led to the development of the Nuremberg Code, which decrees that human test subjects must be knowing volunteers.

The situation is made infinitely worse, however, when the guilty parties attempt to cover their tracks and to hide their evil activity. Not only is their original conduct offensive, but so too is the attempt to hide it or deny responsibility, as this often prolongs the misery and extends the time taken to obtain justifiable compensation for the victims. If vital details about the accidents or wrongdoings are not released speedily, it can even result in the inability to provide those suffering the effects with medical treatment. If you seek to disclaim responsibility for poisoning somebody, it is not in the interest of your defense to declare the nature of the poison.

It is often the case that corporations, anxious to maximize profits to their shareholders, cut corners when it comes to maintaining proper safety standards and environmental policies. This charge is particularly relevant in underdeveloped regions of the world, where controls are lax, public outcry against environmental pollution is curbed and investigators often find it difficult to obtain on-the-spot information. One of the chief complaints against the activities of the Shell subsidiary in Nigeria has been that allegedly the company did not take adequate measures to safeguard the environment in the Niger Delta. It is hard to envisage such shortcomings being tolerated in the leafy suburbs of North America or Western Europe.

In some instances, the violation of human rights under the cover of confidentiality has had a marked and protracted effect on social attitudes. It is a sobering thought that decades after the Tuskegee syphilis experiment in Alabama was closed in 1972, many black Americans are reluctant to participate in medical studies, and a small percentage believe that the HIV virus was spread deliberately among them by the predominantly white U.S. authorities.

Secrecy over scientific experimentation during wartime inevitably provides some of the most obscene illustrations. During World War II, the Nazi authorities in Germany took the position that the inmates of concentration camps were the property of the state and therefore possessed no human rights. This allowed scientists, particularly those in the medical field, to perform wholesale experiments on prisoners without moral restraint and with little thought for the consequences. These people became caught up in a spiral of brutality that, as the war ended, they made desperate attempts to conceal, largely by killing off those survivors who might otherwise be in a position to give evidence on what had taken place behind the barbed wire and electric fences.

But the Nazis were not the only ones who conducted barbaric experiments on human beings during wartime. The Imperial Japanese army was no less guilty of clandestine human rights violations when it set up the infamous Unit 731 near the city of Harbin in northern Manchuria. Ironically, in Britain, where the political establishment and national press condemned Germany's experimentation, the military, with support from the government, used its own soldiers as guinea pigs in studies on chemical weapons without their consent, fooling the men into thinking they were finding a cure for the common cold.

Accurate information is sometimes hard to come by in underdeveloped parts of the world, but accusations have been levelled that a subsidiary of Shell in Nigeria did not take adequate measures to safeguard the environment against oil pollution in the Niger Delta.

7.1 Porton Down

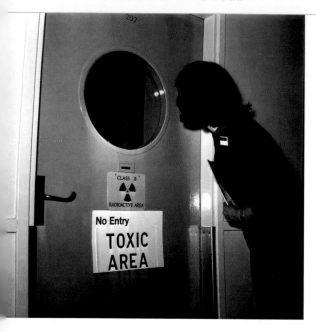

A visitor peeps through the window of a No Entry laboratory at Porton Down, Britain's anti-germ warfare laboratories.

In the decades after World War II, the U.K. weapons research establishment at Porton Down in Wiltshire became synonymous in the eyes of the public with sinister and secretive goings-on. Porton Down was set up shortly after the end of World War I's hostilities in 1918, but maintained a low profile until the 1950s when many former soldiers volunteered as "guinea pigs" for a project that they claim was advertised as research to find a cure for the common cold. It appears, however, they were being recruited for exposure to nerve gases and other chemical and biological agents.

Some of this shadowy history came back under the media spotlight in 2002 when the U.K. attorney general announced his support for a fresh inquest into the suspicious death of Ronald Maddison, a 20-year-old RAF mechanic who had been recruited to Porton Down in 1952. Maddison had been admitted to a special chamber that was then sealed and a drop of the nerve agent sarin (used in the terrorist attack on the Tokyo subway in 1995) was applied to the skin of his arm. The original inquest of 1953 had concluded that Maddison died of asphyxia and it recorded a verdict of misadventure. The inquest was, however, held in secret, on the grounds of "national security." Former volunteers at the facility have indicated that the new inquest would "open a can of worms the likes of which have never been seen before." In April 2002, the U.K. attorney general supported a request by the Wiltshire coroner to open a fresh hearing. A spokeswoman for the attorney general's office said the inquest was to be reopened on grounds of new information not available to the coroner in 1953.

The case for the prosecution

So what did take place at Porton Down and how true are the allegations? One of the major obstacles to getting at the truth is that many of those involved in the experiments during the 1950s and 1960s have since died. A victim who survived the experience, Ray Hutchins, just 18 at the time, alleges that he also thought the experiments were designed to find a cold cure. He was placed in a chamber filled with nerve gas that left him in excruciating pain, but he was terrified of the Ministry of Defence (MOD) and considered that the Official Secrets Act was there to be upheld, so he said nothing.

In 1999, the Wiltshire police started an investigation into the accusations and in August 2001, they sent a dossier of their findings to the Crown Prosecution Service on the grounds that more than 20,000 servicemen appeared to have been subjected to potentially lethal chemicals and other agents. The MOD has denied the allegations, and when the local coroner applied to the High Court for the Maddison inquest verdict to be changed and for a public inquiry to take place, the government rejected his request.

The issue of Porton Down secrecy took on wider significance after a government report was published in the spring of 2002 claiming to give, for the first time, a comprehensive history of Britain's biological weapons trials between 1940 and 1979. Successive governments have tried to keep details of germ warfare tests secret, but the latest document reveals details of more than 100 secret experiments, about which military personnel were instructed to tell the "inquisitive inquirer"

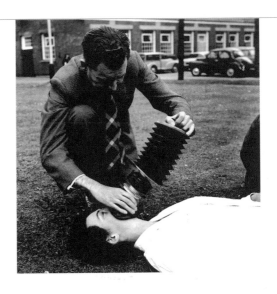

May 23, 1964, Porton Down: Two researchers wearing protective clothing are engaged in producing anthrax for the production of a protective vaccine.

May 23, 1964, Porton Down: A demonstration of one way of treating a nerve agent victim. The man was using an air bellows resuscitator, which was then considered more efficient than the kiss of life.

⬆

The sarin nerve gas attack on the Tokyo subway system in 1995. Sarin was allegedly tested on British military personnel in the 1950s, some of whom died.

that they concerned "weather and pollution." In reality, some of the trials included an unsuspecting general public and were intended to mimic germ warfare by using materials that the MOD claim to have been harmless. One trial required military aircraft to spray zinc cadmium sulphide over large areas, some highly populated, between northeast England and Cornwall.

Did they or didn't they?

According to an *Observer* newspaper report from April 21, 2002, the MOD asserted that "in most cases the trials did not use biological weapons." The wording of the MOD statement suggests, however, that some of the trials did use biological agents and a number of families living close to areas of the country where tests took place now believe that their children have suffered birth defects and physical handicaps as a consequence. Such a possibility is strenuously denied by the MOD. In recent years, the ministry commissioned scientific reports into the safety of the tests and these concluded that there was "no risk to public health." One of the scientists involved in the research conceded, however, that the elderly and people suffering from breathing illnesses may have been "seriously harmed if they inhaled sufficient quantities of micro-organisms."

Liberal Democrat Member of Parliament Norman Baker has posed the question of why it has taken so long to release the information about Porton Down testing over public areas? He has commented, "It is unacceptable that the public were treated as guinea pigs without their knowledge." When asked if such tests were still being carried out, a spokeswoman for Porton Down, Sue Ellison, told the *Observer*, "It is not our policy to discuss ongoing research."

REFERENCES

Barnett, A. Germ warfare tests used on millions. *The Observer.* April 21, 2002

Care, A. Poisoned by their own people. *The Independent.* October 3, 2000.
www.millennium-debate.org/ind3oct.htm

Evans, R. Cold War guinea pigs. *Saga.* September 2001

New inquest on Porton Down victim. *BBC.* April 22nd 2002.
http://news.bbc.co.uk/hi/english/uk/england/newsid_1944000/194442/.stm

Sinclair, K. Porton Down secrets spread to death probe. *The Herald.* August 2, 2000.
www.theherald.co.uk/news/archive/2-8-19100-0-34-41.html

7.2 Unit 731

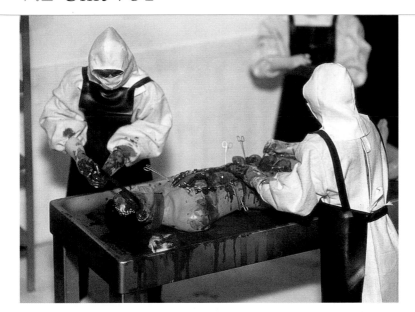

A Japanese exhibition display recreating the atrocities carried out by Unit 731 of the Imperial Japanese Army during World War II. Some of the autopsies were carried out on living victims.

Of all the clandestine activities that went on in contravention of basic human rights during World War II, one of the most infamous was conducted by an elite and specialized section of the Japanese Imperial Army known as Unit 731. This unit, based in Japanese-held Manchuria, performed experiments on thousands of Chinese, Korean and Russian prisoners of war to further the nation's biological weapons program. The emperor of that time, Hirohito, had provided a massive prewar budget to develop the program shortly after Japanese troops invaded Manchuria in 1932.

A physician and army officer named Shiro Ishii, who had graduated from Kyoto University in 1927, headed the program. In 1936, under his command, a huge complex of more than 150 buildings extending over 2.3 square miles (six square kilometers) was constructed at Pingfan, a village close to the city of Harbin in the north of Manchuria. The complex was disguised as a water purification plant and until 1941 was referred to as the "Epidemic Prevention and Water Purification Department of the Kuantung Army." It was, in reality, a germ-warfare research establishment, its mandate being the development of weapons of mass destruction with which to win a future war.

Senator Dianne Feinstein introduced a bill to have Japanese war crimes documents declassified in 1995. Unfortunately, this bill is still ongoing.

The Geneva Protocol of 1925 had banned germ warfare under international law, but the Japanese did not sign it until much later. Perversely, it may have been this international agreement that provided the stimulus for the Japanese policy to develop such formidable weapons. The military leaders who formulated the protocol had concluded that such devices were "too awful to ignore." Once brought to light, the atrocities that were committed in pursuit of the goal of producing biological weapons of mass destruction, in which more than 10,000 POWs were used and then slaughtered in much the same way as laboratory rats, left an indelible legacy of horror and shame.

Human vivisection was a common occurrence at Pingfan.

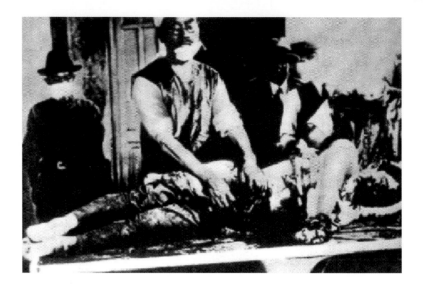

The horrors of war

The materials developed through the research program were not only tested on prisoners of war at Harbin, but also on a much larger scale on Chinese civilians. It is estimated that up to 300,000 people fell victim to Japan's program between 1938 and 1945, when the Japanese military dropped small "bombs" containing biological agents—including bubonic plague (along with the fleas through which the disease is spread), anthrax and cholera—onto unsuspecting sections of the Chinese population. Some of those who fell ill were then rounded up by Japanese soldiers and, allegedly, cut open while still alive so that internal samples could be collected. Similar atrocities are reported to have taken place within the Manchurian facilities of Unit 731.

Some of the most shocking revelations came through a newspaper article entitled "Japan Confronting Gruesome War Atrocity." Compiled by Nicholas D. Kristof, it was published in the *New York Times* on March 17, 1995, and gave a detailed account of what had taken place in Unit 731, including the frequent human vivisection.

An international coverup

An equally severe indictment should also be reserved for the United States. In the autumn of 1945, Japanese troops destroyed the headquarters of Unit 731 and Ishii ordered that the remaining prisoners be killed to cover the evidence. When the U.S. military discovered what had taken place at Unit 731, they realized that not only was it essential to keep the details from falling into the hands of the Soviets, but that the knowledge gained would place them in a favorable position in the arms race.

In 1946, U.S. General MacArthur was given the go-ahead for a secret trade-off wherein the members of Unit 731 would be granted immunity from war crimes prosecution in exchange for the biological warfare research data that they had amassed from their experimentation. The deal was completed in 1948, but soon after the last of the documents was turned over, it was claimed that they were virtually useless. Many of the notes were transferred to microfilm and stored in U.S. military archives before the originals were returned to Japan in 1953.

Some of the details began to emerge in 1985, when a former U.S. army lieutenant colonel Dr. Murray Sanders claimed that, as a military advisor on biological warfare in the autumn of 1945, he had been sent to Japan to investigate Unit 731 and had then persuaded MacArthur of the value in approving an immunity deal. Further light was shed on the issue during an inconclusive one-day investigation by a U.S. congressional subcommittee in 1986, when it was suggested that U.S. servicemen captured in the Philippines and sent to Manchuria might also have been victims of the Unit 731 experiments. In 1989, the first serious exposé was compiled by two British journalists, Peter Williams and David Wallace, in a book entitled *Unit 731: Japan's Secret Biological Warfare in World War II*.

In December 1999, U.S. senator Dianne Feinstein introduced a bill to have the documents relating to Japanese war crimes held by the U.S.

Ruins of the bacteriological factory used by Unit 731 in Ha'erbin.

General Shiro Ishii, the commander of Unit 731.

government declassified and released. These were to cover the period between the Japanese invasion of Manchuria in 1931 and the last year in which Washington received Japanese Imperial Army records. The Bill was reintroduced in both February and October 2000, but was referred to the Government Reform Committee and the Select Intelligence Committee. No vote has been taken.

In 1995, a pressure group made up of Japanese and Chinese members, appalled at the continuing denials and coverups, brought a lawsuit against the Japanese government in an attempt to gain admission of the extent to which Imperial Japan waged biological warfare against the Chinese, as well as an apology and compensation. The project planning began in 1955 and much of the work, including the drumming up of public sentiment, has since been spearheaded by Wang Xuan, whose relatives in the village of Chong Shan were plague-bombed. She has made it her lifetime vocation to bring the Japanese government to account. The verdict was due in the spring of 2002 but, as yet, the decision remains pending.

Bacteriological bombs of the Japanese army.

REFERENCES

Flatin, P. Release of all data on Japan war crimes urged. *Kyodo News.* Through Nichi Bei Times online. December 22, 1999.
www.nichibeitimes.com/news/warcrimes/html

Germ warfare timeline. *Global Alliance for Preserving the History of World War II in Asia.* August 13, 1995.
www.museums.cnd.org/njmassacre/recent-news2.html

Japanese Unit 731: Biological Warfare Unit. *Pacific War, WW2.* January 21, 2001.
www.marchallnet.com/~manor/ww2/unit731.html

Unit 731. *BBC News Online.* February 1, 2002.
http://news.bbc.co.uk/hi/english/audiovideo/programs/ correspondent/ne.../1796044.st

Wu, Tien-wei. A preliminary review of studies of Japanese biological warfare and Unit 731 in the United States. *Centurychina.* Undated.
www.centurychina.com/wiihist/germwar/731rev.htm

Bills introduced in the 106th Congress. Updated January 6, 2001.
http://www.cohnmarks.com/LegislUpdate/106thBills.htm

7.3 Shell in the Niger Delta

Kolo, Niger Delta. Young activists take over a Shell-owned flow station. The company allegedly wreaked havoc on the local environment by using outmoded techniques and equipment.

In 2001, Harry Roels, on the Committee of Managing Directors of the Shell International Petroleum Company, declared, "We've taken the ideas of transparency and openness fully on board—even when it is difficult. I'm not sure how this will play out but a lot of people in Shell—particularly younger people—are very pleased with it." How many were "very pleased" and how many viewed these statements with a measure of cynicism was not made public, since some would argue that transparency and openness have not been among Shell's strongest attributes. It has been suggested that while Shell, in common with other oil companies, demonstrates responsibility over the social and environmental impact of its activities in the developed world, this conscience does not always extend to third world regions where the media pay little attention, public accountability does not take center stage and access for independent verification is fraught with difficulty.

Less well highlighted is a recently published comment by Shell on the activities of its subsidiary in Nigeria, the Shell Petroleum Development Company (SPDC). "There have been claims that SPDC's operations have devastated the environment. Indeed, there are undeniable environmental problems in the Niger Delta and it is equally true that the oil industry has contributed to these. But they do not add up to 'environmental devastation'." Some would see this as a gross understatement of a situation that Shell would have preferred to keep hidden from the outside world.

Ken Saro-Wiwa, who spoke out against decades of environmental abuse in Nigeria. He was imprisoned for over 17 months and finally sentenced to death by hanging.

More than 150,000 barrels of oil have spilled into the Niger Delta. Local Ogoni people who offer resistance have been persecuted by the Nigerian government.

What has Shell done?

SPDC activities have severely affected the lives of the Ogoni people in the Delta region, where much of Shell's oil exploration and drilling takes place as a joint venture with the state-owned Nigerian National Petroleum Corporation (NNPC). Shell has been accused of poor environmental management and collusion with the Nigerian military authorities in human rights violations—allegations that it refutes. Former head of Public Affairs for Shell Richard Tookey has stated that SPDC always tries to minimize the impact of operations on the environment and to ensure that local communities gain real benefits from having a Shell company as a neighbor.

Local reports from the Niger Delta speak of havoc being wreaked on communities and the environment because the company employs operational techniques and equipment so outmoded that they would be considered illegal in other parts of the world. The Ogoni people, who have traditionally resisted the exploitation of their homeland, have been persecuted by the Nigerian government with methods that have included beating, jailing and, on occasion, fatal shooting by security forces. Shell, it is said, has done little to curb these excesses and has turned a deaf ear to the entreaties of human rights organizations.

The company's attitude

Critics accuse the company of adopting cavalier attitudes toward the local environment, with a notable absence of the kind of scrupulous assessments that would be undertaken if similar projects were in North America or Western Europe. When Shell laid a pipeline from Cheshire to Scotland, for example, it cataloged every yard of the route and replaced each hedge, wall and fence exactly as it had been before the company arrived, "to avoid lasting disfiguration and accommodate environmental concerns." This stands in stark contrast to the company's policy in Nigeria. It has been alleged that at no time did SPDC consult the Ogoni people about the environmental impact of its drilling, nor has it shown them an assessment. Pipelines have been installed that run through the middle of villages and across essential farmland, making it virtually useless. A similar story is told over the siting of flares that burn off surplus gases, an issue that particularly vexes local residents. Some of these flares have run night and day, year after year, within 330 feet (100 meters) of homes, a practice that is said to cause respiratory diseases and perennial contamination with soot. The late Ken Saro-Wiwa, spokesman for the Movement for the Survival of the Ogoni People (MOSOP), revealed that flares had been installed dangerously close to several villages in the Yorla and Korokoro oilfields.

Oil spillage is also reported in the Niger Delta at levels that would be considered intolerable in areas where media attention is better focused and accountability is a more serious factor. An independent report indicates that in the decade following 1982, Shell's Nigerian activities

Ogoniland, Nigeria: Burning oil
from a leaking pipeline.

discharged more than 1.5 million gallons of oil into the local environment through accidental spillage. Shell claims that much of the problem is caused by sabotage, but this has been strongly denied by a spokesman for the UN World Commission on Development and Culture who described it as "the kind of irresponsible propaganda that the oil companies are putting out in order to discredit those who are trying to do something about the environment." Substantiation of Shell's claim has been extremely difficult because of considerable obstacles to independent verification.

Money, money, money

Both Shell and the Nigerian government have earned substantial income from oil drilling activities in the Niger Delta. For a 25-year period, the figure has been put at around $30 billion dollars worth of extraction in the Ogoni region alone. During the same period of time, it is estimated that Shell provided a mere $200,000 of local community assistance, and Ogoni representatives allege that Shell has provided no compensation for the loss of farmland, only for loss of crops. A leaked internal report prepared by independent consultants reveals that "SPDC development activity remains 'top down' and consultation extends only as far as the local elite...classrooms are built but not furnished, a commuter bus sits locked away, transformers have burned out, drugs are dispensed without medical supervision."

One of the most serious complaints, however, concerns violations of human rights. Using the defense that its employees need protection from hostile local activists, and in common with other oil companies in

the region, Shell has employed security forces. There have been repeated accusations that the company revokes its responsibility to ensure that these armed policemen act in a proper and humanitarian manner.

In fairness, after Shell's Nigerian activities came under scrutiny in 1995, it undertook a major review of its policy over local community issues and human rights. In July 2001, Ron M. van den Berg, the Managing Director of SPDC, was summoned before the Human Rights Violations Investigation Commission to explain the conduct of his company towards the Ogoni people. He admitted that aspects of his company's operations had attracted adverse attention, but excused gas flaring on account of "limited opportunities for its utilization" and avoided the matter of siting. Van den Berg reiterated Shell's wholly unsubstantiated argument that oil spillage was largely due to "criminal activities," yet avoided the human rights violation question and protested that his company is recognized for its "openness and transparency."

Shell is now committed to environmental improvement but questions remain. Would it have happened without exposure and pressure from outside and how strong is the commitment? In his closing statement to the commission, Van den Berg conceded: "It is SPDC's hope that agreement can be reached with MOSOP and the Ogoni people...so that lasting peace and reconciliation may be achieved." Clearly there is still some way to go.

The Ogoni people, who live in the Niger Delta region of Nigeria, have long held demonstrations against Shell, but to little avail. Today, international pressure groups and charities help to make this a worldwide campaign.

REFERENCES

Controversies affecting Shell in Nigeria. *Report to Clients. PIRC.* March 1996.
www.pirc.co.uk/shellmar/htm

The price of oil. Corporate responsibility and human rights violations in Nigeria's oil producing communities. *Human Rights Watch.* January 1999.
www.hrw.org/reports/1999/nigeria/index/htm

Rowell, A. (ed. Goodall, A.). The environmental and social costs of living next door to Shell. *Greenpeace.* July 1994.
www.greenpeace.org/~comms/ken/enviro.html

Shell Petroleum Development Company of Nigeria.
Shell Nigeria. Undated.
www.shellnigeria.com/

Shell Petroleum Development Company of Nigeria. Factfile.
Shell Nigeria. Undated.
www.shellnigeria.com/frame.asp?Page=factfile

Shell Petroleum Development Company of Nigeria. A statement presented by Mr. Ron M. van den Berg Managing Director . . . to the Human Rights Violations Investigation Commission.
Shell Nigeria. July, 24, 2001.
www.shellnigeria.com/frame.asp?Page=info

Shell's Nigeria development projects slammed in internal report.
AfricaOnline. September 27, 2001.
www.africaonline.com/site/Articles/1%2C3%2C41103.jsp

7.4 The Tuskegee Syphilis Study

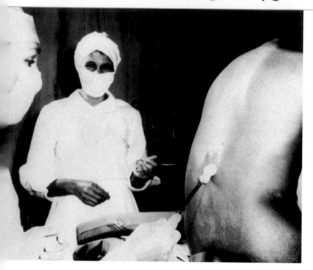

Nurse Eunice Rivers assisting U.S. Public Health Service doctor as he does a spinal tap on a patient.

In the summer of 1972, Associated Press journalist Jean Heller compiled a report for the *Washington Star* exposing one of the greatest scandals ever to take place in the name of medical science. On July 25 of that year, the newspaper published an interview given to Heller by a former employee of the United States Public Health Service (PHS), Peter Buxtun. It revealed details of a highly secretive experiment run by the PHS in cooperation with the Tuskegee Institute, a black university whose affiliated hospital had lent its medical facilities and its name to the study. The program, in operation since 1932, targeted 399 black men, mostly illiterate sharecroppers living in Macon County, one of the poorest areas of the southern state of Alabama and, in the pithy words of news anchorman Harry Reasoner, "used human beings as laboratory animals in a long and inefficient study of how long it takes to kill someone with syphilis."

What is syphilis?

Syphilis, caused by a microorganism called Treponema pallidum that is passed on during sexual contact, includes three stages: it begins with ulcers, then after some weeks or months a rash similar to that of chicken pox appears, followed in the advanced stages by tumors, heart problems, paralysis, blindness, insanity and finally death. Syphilis can occur in several forms and in one, neurosyphilis, the organism invades the cerebrospinal fluid. This variation was, allegedly, the justification for the research since it was believed in some medical circles that white people suffering from syphilis experienced greater neurological complications, while black people were particularly prone to the cardiovascular ailments. To put the theory to the test, the PHS needed to observe the course of the disease through all its stages. In order to gain community support, the service devised a program, trumpeted at the time as one of great scientific merit, whereby patients were led to believe that they were receiving the best medical attention; in reality, the disease was being left to run its course. The mantra taught to nurses working on the project was "never diagnose, never prescribe, follow doctor's instructions."

By the time the study was closed in 1972 amid a wave of public outrage, it had directly and indirectly killed 128 people, and over the course of time it would lead to the premature death of nearly 400 male subjects and their relatives. Yet when the story had first broken, an Alabama state health official brushed it aside with the claim that "someone is trying to make a mountain out of a molehill," and for many years the PHS remained unrepentant. It insisted that the victims had been "volunteers" and individual doctors offered the same morally corrupt defense that had been presented in the postwar Nuremberg trials—they were "just carrying out orders." Critics quickly drew comparisons with Nazi Germany. Arthur Caplan, Director of Medical Ethics at the University of Pennsylvania, was one of the first to describe what had taken place as "America's Nuremberg."

It seems extraordinary that it took more than 30 years of public pressure to earn the victims an official apology, but not until May 16, 1997, did the eight survivors of the Tuskegee experiment hear President Clinton state:

The United States government did something that was wrong— deeply, profoundly, morally wrong. It was an outrage to our commitment to integrity and equality for all our citizens...what was done cannot be undone but we can end the silence. We can stop turning our heads away. We can look at you in the eye and finally say, on behalf of the American people, what the United States government did was shameful and I am sorry.

Setting ethics aside

The principle of the Tuskegee study amounted to an epitome of bad science and even today there is no coherent explanation for the benefits that might have resulted. Impoverished but trusting men suffering from syphilis, most of whom had probably never previously seen a doctor in their lives, were to be offered the lure of free meals, cash handouts and free medical care to treat them for what was euphemistically described as "bad blood." Treatment and cure for these casualties of ethical madness was not, however, on the agenda of the PHS, rather the reverse. As one commentator put it, Tuskegee was "the longest non-therapeutic experiment on human beings in medical history."

When penicillin became widely available in 1943, the administrators of the study needed to be especially zealous in keeping the men away

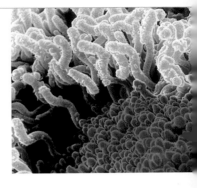

Color scanning micrograph of a colony of bacteria *Treponema*. Pathogenic species of *Treponema* include *T. pallidum* which causes syphilis. In syphilis, the bacteria may be found on the duodenum (the first part of the small intestine), as seen here, as well as in most other parts of the body.

May 1997: U.S. President Bill Clinton apologizes publicly to victims of the Tuskegee syphilis experiment.

from effective treatment. In spite of various local campaigns to eradicate venereal disease, Macon County doctors, VD clinics and military draft boards were provided with lists of the subjects to ensure that all patients were barred from receiving the antibiotic medication that would have offered them a way back to health.

As a public relations exercise and to maintain trust, the patients were, from the outset, dispensed simple iron supplements and pills called "pink medicine." Yet these contained nothing more therapeutic than aspirin. Monitoring the course of the illness also required nurses to take regular blood samples, adding to the patients' illusion that something was being done. Those suffering from the neurological form of the disease were subjected to extremely painful lumbar punctures to draw off spinal fluid. To make sure that they turned up for this dangerous procedure, each patient received an encouraging letter promising a "last chance to get special free treatment."

What none of the recipients of the "free treatment" knew was that they were actually on a one-way ticket to the mortuary and that their remains would receive postmortem autopsies. This was a particularly sensitive issue that required the maximum amount of discretion on the part of the Tuskegee teams, since to some people, the idea of one's body being chopped up after death is horrific and carries all kinds of superstitious connotations. As a doctor at the time succinctly put it: "If the colored population becomes aware that accepting free hospital care means a postmortem, every darkie will leave Macon County."

Long-term damage to public trust

It was only after Jean Heller had published her interview with Buxtun, the "whistle-blower," that the sparse handful of survivors of the

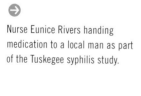

Nurse Eunice Rivers handing medication to a local man as part of the Tuskegee syphilis study.

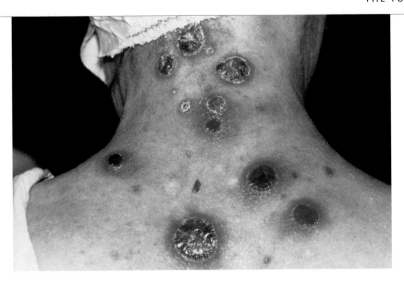

Secondary syphilis rash and inflammation on a back. Syphilis is caused by the *Treponema pallidum* bacterium. It is transmitted through sexual intercourse with an infected person. About six to 12 weeks after infection, secondary syphilis occurs. The main symptom is a skin rash, with headache, fatigue and fever. Tertiary syphilis follows within 10 years of infection, with the formation of tumour-like gummas, causing neurological disorders. Treatment is with antibiotic drugs.

experiment were given proper medical treatment and eventually received the benefit of a $10 million compensation package. But the Tuskegee deception has left a chilling epitaph. A 1990 survey found that 10 percent of black Americans firmly believed that AIDS had been developed by the United States government in order to exterminate blacks. A further 20 percent considered the scenario to be a distinct possibility. Today, many black people in the U.S. shun clinical trials and such vital programs as genetic cancer studies.

REFERENCES

Coming to terms with the legacy of the Tuskegee Syphilis Study. *Tuskegee symposium CMHSL.* February 23, 1994.
www.med.virginia.edu/hs-library/historical/apology/symp.html

How Tuskegee changed research practices. *National Center for HIV, STD and TB Prevention.* April 29, 2002.
www.cdc.gov/nchstp/od/tuskegee/after.htm

The Presidential Apology. Undated.
www.med.virginia.edu/hs-library/historical/apology/whouse.html.

Pyle, K. C. Objective article: Tuskegee Syphilis Experiment. Undated.
www.dreamscape.com/morgana/adrastea.htm

Syphilis Information. *National Center for HIV, STD and TB Prevention.* April 29, 2002 (last reviewed June 19, 2001).
www.cdc.gov/nchstp/od/tuskegee/syphilis.htm.

The troubling legacy of the Tuskegee syphilis study. Historical Collections. *University of Virginia Health System.* Undated.
www.med.virginia.edu/hs-library/historical/apology/html

The Tuskegee Syphilis Experiment. *Infoplease.com.* 2002.
www.infoplease.com/ipa/A0762136.html.

The Tuskegee Syphilis Study legacy committee report. May 20, 1996.
www.med.virginia.edu/hs-library/historical/apology/report.html

Wolinsky, H. Steps still being taken to undo damage of "America's Nuremberg." *Annals of Internal Medicine (American College of Physicians American Society of Internal Medicine).* August 15, 1997.
www.acponline.org/journals/annals.15aug97/currnazi.htm

On October 10, 2000, a lawsuit was filed in Philadelphia, Pennsylvania, on behalf of 298 former inmates of the city's Holmesburg jail. The prison was closed in 1995, but it was alleged that prior to that time, prisoners — mostly black males with poor education — had been subjected to "human guinea pig" medical experiments. These took place over a 23-year period from 1951 and were stopped only in 1974 after revelatory congressional hearings into medical experiments elsewhere in the United States, in which individuals were said to have been coerced unwittingly into participation. The most widely reported of these cases was the long-term study conducted by the Tuskegee Institute in Alabama on black men infected with syphilis.

What was tested?

The Holmesburg experiments, said by former prisoner Edward Anthony to have taken place at a special prison block, involved exposure of some of the prison inmates to doses of radiation, carcinogenic pesticides such as dioxin, psychotropic drugs and an assortment of infectious diseases. These experiments, the inmates claimed, had been carried out without obtaining their knowing consent. In other words, they agreed to be the subjects but were not informed about the medical nature of the tests. Claims were put forward that a high proportion of the men had succumbed to cancers, respiratory problems and an assortment of other diseases. During a CNN interview, Anthony displayed badly deformed fingers and alleged that his hands had swollen "as big as boxing gloves"

The medical administrator of the Holmesburg research programme with an inmate undergoing surface patch tests in 1966.

during tests that were performed on him using bubble-bath ingredients. The prison's facilities became well known among the scientific community, not least because of the dermatological research by Dr. Albert Kligman, an emeritus professor and research dermatologist working with the University of Pennsylvania, who is credited for developing an anti-wrinkle and acne treatment known as Retin-A.

The alternatives to "volunteering" for human experimentation were not very appealing. The few jobs in clothing and shoe factories on site were badly paid and required long shifts to be worked.

The trial

Defendants include the Dow Chemical Company and the pharmaceutical giant Johnson & Johnson, each of whose products were allegedly used on the inmates. Other defendants are Dr. Kligman and the City of Philadelphia authorities. The lawsuit, presented in the Philadelphia Common Pleas Court, demanded $50,000 in damages from each of the defendants, accusing them of "negligence, carelessness and recklessness."

According to the prosecution attorney Thomas Nocella, the Holmesburg inmates had not been given any information about the nature of the chemicals and drugs to which they were being exposed and could not, therefore, have given informed consent. They were paid a few dollars a day for their cooperation. By contrast, the prison received annual fees from various companies in the order of "hundreds of thousands of dollars." Kligman and Johnson & Johnson benefited financially from Retin-A income, and the university gained when it sued and won a percentage of the Retin-A profits. The university and the city had themselves been sued for sums of up to $40,000 by some of the inmates as early as 1984.

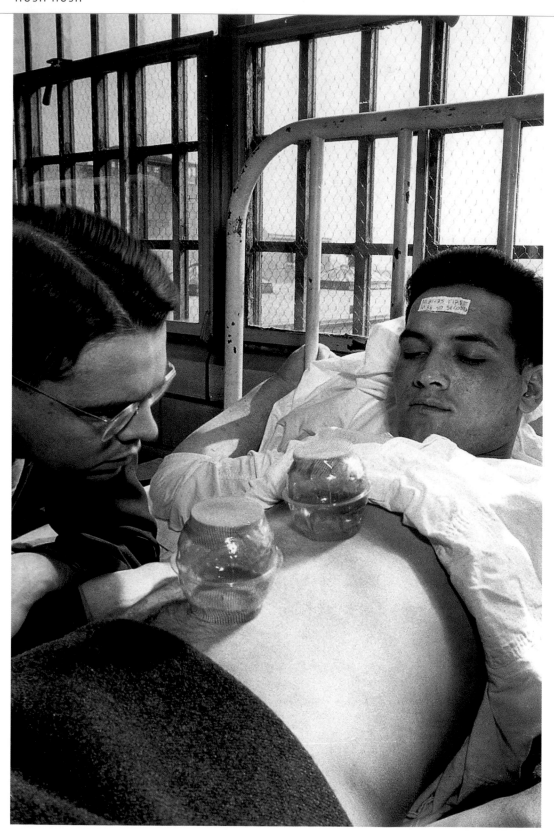

The Holmesburg situation was fully exposed by a former teacher of literacy at the prison, Allen Hornblum, in his book *Acres of Skin*, published in 1998. Hornblum reported that he had seen many inmates with bandages on various parts of their body and assumed at first that these were evidence of fights. Only later did he discover that they were the more visible signs of the medical experiments. In November 1998, former inmates organized a protest march to the university hospital to draw attention to their plight and to demand compensation. They alleged that they had been "lied to, used and exploited as human guinea pigs." Later that year, a spokesman for the university, Richard Tannen, suggested that the university hospital provide free medical evaluation for inmates who had taken part in the tests but would not guarantee free treatment.

When the lawsuit was filed in 2000, a spokeswoman for the University of Pennsylvania, Rebecca Harmon, claimed that although the use of prison inmates for medical experimentation had been common practice during the 1950s and 1960s, the policy of the university had now been changed and such experimentation was discouraged.

John Stanley, of the University of Pennsylvania Department of Dermatology, put forward the familiar academic response that Hornblum had misinterpreted the nature of the experiments, possibly due to the "lack of experience of the lay author." His choice of words is perhaps significant when, on the one hand, he states that "there is no evidence, or even any suggestion, that any prisoner was killed by any experiment," while on the other he offered the more reserved comment "it is not clear that any prisoner was harmed by the studies at Holmesburg." Put another way, Stanley seems to have been conceding that there was no evidence that inmates had not been harmed.

The spoke and wheel design of Holmesburg prison, built in 1896 and closed in 1995. It was the focus of human guinea pig medical experiments on inmates between 1951 and 1974.

Holmesburg was not the only U.S. prison to conduct studies on inmates. This man was an inmate at Stateville Penitentiary, Illinois; he was being bitten by mosquitoes carrying malaria so that scientists could research the disease.

REFERENCES

Dale, M. Inmates sue over medical studies. *Associated Press.* October 18, 2000.
www.abcnews.go.com/sections/us/DailNews/inmates001019.html

Ex-inmates seek treatment for prison experiments. *CNN.* November 8, 1998.
www.cnn.com/us/9811/08/acres.of.skin/

Experimental Research. Ex-inmates sue Penn and Kligman over research. *Pennsylvania Gazette.* February 1, 2001.
www.upenn.edu/gazette/0101/0101/gaz3.html

Krivo, J. Ethics, the prison system and dermatology. *Archives of Dermatology.* April 1999.
http://archderm.ama-assn.org/issues/v135n4/ffull/dlt0499-1.html

Stanley, J. R. Ethical accusations: the loss of common sense. *Archives of Dermatology.* February 2000.
http://archderm.ama-assn.org/issues/v135n2/ffull/dlt0200-4.html

"He lied to parents, he lied to other doctors, lied to hospital managers, he stole medical records, he falsified statistics and reports, and he encouraged other staff to do the same." These would be forthright accusations in any context, yet they are especially noteworthy because they were made by the British secretary of state for health Alan Milburn about Professor Dick van Velzen, who was formerly employed jointly by Liverpool University and the city's Alder Hey Children's Hospital where he was senior pathologist from 1988 until his suspension in 1995.

Professor Dick van Velzen, pathologist, who retained deceased babies' organs without the consent of their next of kin in Alder Hey Hospital, Liverpool, U.K.

The rights of a child, the rights of a human

An investigation revealed that during the time Van Velzen was in charge of pathology at the hospital, the organs of 893 dead children had been removed without the knowledge or consent of their parents. This revelation was only the start of what was to unfold into a shocking catalog of unauthorized organ retention that had begun in 1948. The hospital later admitted retaining the hearts of 2,087 children, even before the discovery that other organs had been stripped. Some of these organs were "stockpiled" in a laboratory owned by the University of Liverpool.

In one of the worst cases to come to light, the organs, including heart, brain and lungs, of a 10-day-old baby who died in 1992 were apparently lost by the hospital and, in consequence, his family was not able to take them for the funeral.

At the end of the 1990s, Alder Hey became the focus of what was seen by many to be an ethical outrage. The first hints that something was amiss in the conduct of the Royal Liverpool Children's National Health Service (NHS) Trust, Alder Hey, emerged during an inquiry into baby deaths at another hospital, Bristol Royal Infirmary. On September 7, 1999, Professor Robert Anderson of the British Paediatric Cardiac Association reported almost in passing that Alder Hey had "probably the biggest and best collection" of hearts in the country. On December 3, in the face of mounting pressure from relatives of children whose organs had been removed without consent, Milburn announced an independent inquiry. The matter had been brought into sharp focus after the Liverpool coroner opened an inquest into the death of a baby, Kayleigh Valentine, whose organs had been taken without parental agreement. The inquiry panel was established in December 1999 under the chairmanship of Michael Redfern, QC.

Twenty-one of the babies whose organs were retained at Alder Hey Hospital after their deaths, without permission from their parents.

The inquiry's conclusions

In its findings, the inquiry panel identified a damning series of management blunders at Alder Hey. Proper financial and material resources had not been provided for pathology. Alder Hey and the

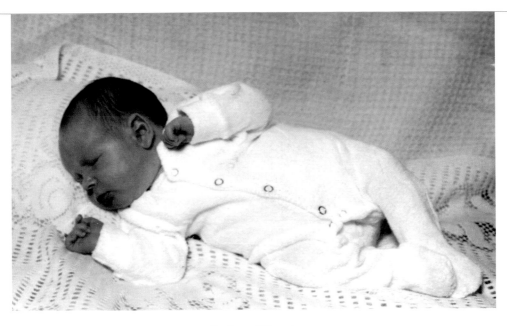

Andrew O'Leary

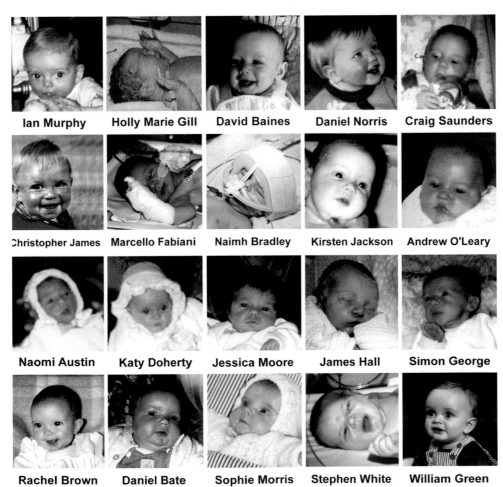

Ian Murphy   Holly Marie Gill   David Baines   Daniel Norris   Craig Saunders

Christopher James   Marcello Fabiani   Naimh Bradley   Kirsten Jackson   Andrew O'Leary

Naomi Austin   Katy Doherty   Jessica Moore   James Hall   Simon George

Rachel Brown   Daniel Bate   Sophie Morris   Stephen White   William Green

Alder Hey Hospital, Liverpool, U.K.

University of Liverpool had permitted Van Velzen to abdicate his clinical duties and responsibilities and had failed to investigate his postmortem practice between 1988 and 1995, which would have revealed the retention of every organ in every case. Senior officials were complicit in Van Velzen's falsifications, and the service manager at Alder Hey had "allowed himself to be sidelined."

On March 16, 2000, the chairman of the trust running the hospital, Frank Taylor, resigned from the board after demands from Alan Milburn. Eight days later, Alder Hey's chief executive Hilary Rowlands went on extended leave. Three other NHS staff members were suspended. On September 30, Canadian police issued a warrant for Van Velzen's arrest after the discovery of child body parts in Nova Scotia. By January 2001 it was apparent that up to 400 complete fetuses had been retained without parental knowledge or consent and the organs of a total of 3,500 children had been removed without such knowledge or consent.

The official inquiry, which was published in the House of Commons at the end of January 2001, reported that Van Velzen had falsified records, statistics, research applications and postmortem reports while at the same time lying to parents about postmortem findings. In some instances, he had even failed to carry out postmortems properly, and he had left his department with a budget deficit of more than £70,000 ($1120,000). The inquiry found that fault extended beyond Van Velzen and that abuse of the system, which began before his arrival at Alder Hey, had been made possible through management failings. Van Velzen had encouraged staff to falsify reports and therefore was not alone in his actions.

Elsewhere

Alder Hey, it has transpired, was also not alone in its conduct over organ removal from children. The BBC conducted an independent survey and found that, until about 1999, seven of the ten biggest NHS trusts in England were removing organs or tissue samples without parental agreement. NHS workers at one hospital, the King's Mill Centre in Nottinghamshire, revealed that they had been instructed to dispose of fetuses, babies' brains, hearts and other organs in waste bags. In January 2001, the BBC reported that the Birmingham Children's Hospital had admitted carrying out organ removals in a manner comparable to that at Alder Hey, although officials confirmed that this was no longer their practice.

The room in Alder Hey Hospital where the babies' organs were stored. The red circle indicates the shelf on which they were kept.

REFERENCES

Alder Hey-the timetable. *BBC News Online*. January 30, 2001.
http://news.bbc.co.uk/hi/english/health/newsid_1144000/1144035.stm

Body parts hospital also hoarded fetuses. *Medserv Medical News*. November 16, 2000.
www.medserv.dk/health/2000/11/16/story01.htm

Details given on organ removal inquiry. *BBC News Online*. December 17, 1999.
http://news.bbc.co.uk/low/english/health/newsid_570000/570452.stm

New organ scandal forces hospital chief to quit. *The Guardian*. March 17, 2000.
www.guardian.co.uk/print/0%2C3858%2C3975030%2c00.html

Pressure piles on Alder Hey. *BBC News Online*. January 26, 2001.
http://news.bbc.co.uk/hi/english/health/newsid_1138000/1138371.stm

Probe into new baby organ scandal. *BBC News Online*. June 4, 2000.
http://news.bbc.co.uk/hi/english/health/newsid_776000/776426.stm

The report of the Royal Liverpool Children's Inquiry. *The Royal Liverpool Children's Inquiry*. January 30, 2001.
www.rlcinquiry.org.uk/

Summary: Management failings. *The Royal Liverpool Children's Inquiry*. January 30, 2001.
www.rlcinquiry.org.uk/summary/sum7.htm

Summary: Organ retention 1948-1988. *The Royal Liverpool Children's Inquiry*. January 30, 2001.
www.rlcinquiry.org.uk/summary/sum4.htm

Ven Velzen's "worst excesses." *BBC News Online*. January 30, 2001.
http://newssearch.bbc.co.uk/hi/english/health/newsid_1144000/1144363.stm

What Does the Future Hold?

A discussion of what happens tomorrow and the next day is perhaps a little pre-emptive. But history has shown us that secrecy is rife, so it is no great assumption that there are secrets being kept right this minute. Secrecy, by its very nature, involves information being withheld. As pointed out in the introduction, a secret is something known to the few, perhaps even a single individual, but is not conveyed to the rest. So you and I learn about such things only because, by design or accident, the cat has been let out of the bag.

The use of genetic engineering to design babies is still largely a thing of the future, but will our desire to tamper with nature bring disastrous consequences?

The pages of this book contain case studies of activities, incidents and situations that have been hushed-up, sometimes deliberately kept under wraps for long periods, but which have eventually been disclosed, brought into the open forum and discussed in the press and on radio and television. Sometimes, the world at large has learned of situations only when legal suits have been filed against alleged culprits or when a damaging exposé appears in a book. Other times, it has come down to public outrage about a scandal over which the public has gradually become acquainted through some other means.

It would be thoroughly naive of us to believe, as we read the ensuing pages, that the coverups and deceits described—many of which have carried the potential to affect us adversely, some of which have done and indeed continue to do so—always reflect conduct of the past. Unfortunately, the world has not suddenly become a more ethical and open place where we can be confident that we are being informed of each and every activity that we would wish to know about for our own well-being. Right now, even in our world of instant telecommunication, we can be reasonably confident that governments are withholding information from us on what they regard to be a need-to-know basis, as are the military, medical research establishments, industrial corporations and others. Secrets being hatched or prolonged today by the few may affect the rest of us tomorrow, but in the meantime we remain blissfully ignorant of what is going on around us. That is, of course, until the next catastrophe-in-waiting takes place or someone uncovers another corporate conspiracy gone wrong, a new medical scandal, a fresh piece of hush-hush. This conclusion is unlike any of the preceding chapters because its scenario is guesswork. It is about things that might be revealed at some time in the future, not what we can examine and criticize in the safety of hindsight today.

There are some areas of covert activity about which we can make informed guesses. We can be reasonably sure that military authorities around the world are busy developing new and more deadly weapons away from our inquisitive eyes. Multinational corporations are hard at work designing next year's cars, fresh drug therapies, revolutionary electronics and so on, while keeping them strictly under wraps until the appointed time when they are launched upon us. Some of these things, if history is to be any judge, will hold the seeds of personal or larger scale disaster.

Medical experiments and clinical practices will go horribly wrong and the outcome will be kept from us for as long as possible in order to protect the perpetrators. There will be terrible accidents, some of which subsequent investigation will no doubt attribute to corporations. Claim and counterclaim about such vital interests as global warming and depletion of resources will leave us little the wiser.

One interesting question concerns if and when we will be told about impending global catastrophe. In many respects, we have been lucky thus far in terms of our health and well-being. We have been beset by AIDS and, recently, by an anthrax scare. Both contain a frightening potential but are, by comparison with some human afflictions, difficult to transmit. Even with the prospect of a terrorist attack involving smallpox or bubonic plague, we are told that at least some national authorities possess adequate stocks of vaccine and antibiotics. But what if a mutant virus or resistant bacterium were to appear that was passed freely as an airborne contaminant and for which we have no available response? How ready would governments be to tell us the truth in such doomsday circumstances, and would those in privileged positions keep the information to themselves as a means of self-protection?

Controversy continues to surround the human form of mad cow disease, BSE. Will the truth come too late?

Theories, some believable, some fantastic, abound about the origins of HIV and Aids. Is what we are being told the truth?

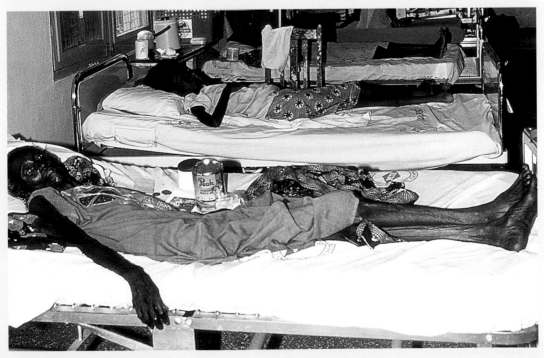

Dolly the sheep, the world's first cloned mammal, with Dr Ian Wilmut who developed this cloning process. At first, Dolly was perfectly healthy, but as time has gone on, she has developed serious arthritis. Are her health problems side-effects of her unusual conception?

We now know, for example, that during the Cold War, most government, military and major corporate authorities had contingency plans to protect themselves in well-stocked underground bunkers, the existence of which has now been revealed, but at the time was not disclosed to the public. The authorities went to considerable lengths to keep it quiet. Such powerful minorities would probably not be at all keen on disclosing the details of impending future catastrophe and the arguments are not hard to anticipate. We have seen them projected luridly in Hollywood disaster movies where the Earth is about to be devastated in one or another form of Armageddon. It is deemed advisable that information is withheld from the public until a point is reached when there is no choice because the threat has become self-evident. Why? Because broadcast of such classified information is likely to result in mass panic and the breakdown, in the few hours or days that are left, of law and order, something that seems to worry our leaders even when we are confronted with the hypothetical prospect of mass oblivion.

Modern science has shown us that we are at risk of global catastrophe from a number of origins that are not of man's making. Geology and paleontology reveal that at times in the distant past, super-volcanoes have devastated large parts of the Earth with magnitudes far in excess of anything that Etna, Vesuvius and other potential hot-spots can throw at us today. Records also reveal that asteroids have periodically struck the Earth with devastating effect, producing vast dust clouds that have blotted out the sunlight and caused mass extinction. The question is, in the event that seismologists, vulcanologists or astronomers should detect such a global catastrophe to be imminent, will you and I be the last to know?

The long-term consequences of genetically-modifying foods are still unknown. Anti-GM pressure groups promote their theories, while the companies who profit from these foods insist on their safety. Only time will tell which side has been correct.

Glossary

acid rain
Rain with a high concentration of acids produced by sulfur dioxide, nitrogen dioxide, etc. resulting from the combustion of fossil fuels.

acrylamide
An organic solid of white, odorless, flake-like crystals. Its greatest use is as a softening aid in drinking water treatment.

ADD/ADHD
Attention deficit disorder (ADD) and attention deficit hyperactivity disorder (ADHD) are diagnostic labels given to people who have problems with inattention, impulsivity, hyperactivity and boredom.

adenoids
Growths of lymphoid tissue in the upper part of the throat.

anthrax
An infectious disease of wild and domesticated animals, which can be transmitted to people. It is characterized by black pustules.

antimicrobials
Able to inhibit or control microbes.

asbestosis
A form of lung disease caused by inhaling asbestos particles.

asphyxia
Loss of consciousness as a result of too little oxygen and too much carbon dioxide in the blood; suffocation causes asphyxia.

biotechnology
The use of engineering and technology on living organisms or their components, such as enzymes, to make products that include wine, cheese, beer and yogurt.

biological warfare
The deliberate use of disease-spreading microorganisms, toxins, etc. in warfare.

bubonic plague
A contagious disease, the most common form of plague, caused by a bacterium transmitted by fleas from infected rats. It is characterized by buboes, fever, prostration and delirium.

carbon dioxide
A colorless, odorless, incombustible gas, heavier than air, that is a product of respiration and combustion.

carcinogen
Any substance that causes cancer.

chemical warfare
Warfare by means of chemicals and chemical devices such as poisonous gases, incendiary bombs, smoke screens, etc.

chlorpyrifos
A broad-spectrum organophosphate insecticide.

cholera
Any of several intestinal diseases.

deep vein thrombosis (DVT)
The formation of a blood clot in one of the deep veins of the body, usually in the leg.

defoliant
A chemical substance that causes leaves to fall from growing plants.

depleted uranium (DU)
The uranium remaining after that used for nuclear weapons and reactors has been extracted from natural uranium; highly toxic and accounting for 99% of uranium.

dichlorophenoxyacetic acid
The first successful selective herbicide developed.

dioxin
Any of a number of mostly poisonous chemical byproducts of the manufacture of certain herbicides and bactericides, including TCDD, in particular its variant of trichlorophenoxyacetic acid, a component of Agent Orange.

DNA
Deoxyribonucleic acid; a nucleic acid that is the main constituent of the chromosomes of all organisms (except some viruses). DNA is self-replicating, plays a central role in protein synthesis and is responsible for the transmission of hereditary characteristics.

electromagnetic (ray)
Of, produced by or having to do with electromagnetism or an electromagnet.

epidemiology
Branch of medicine that investigates epidemics.

exothermic (reaction)
A chemical change in which there is a liberation of heat, as in combustion.

fluoridate
To add fluorides to (a supply of drinking water) in order to reduce the incidence of tooth decay.

fluorosis
A disorder resulting from the absorption of too much fluorine.

genetically modified
Technologies that alter the genetic makeup of such living organisms as animals, plants or bacteria.

germ-weapons
Living organisms adapted for military use and intended to cause disease or death in human, animal or plant life.

greenhouse effect (gases)
The heating of the planet by carbon dioxide and other gases in the atmosphere absorbing infrared radiation emitted by the planet as a result of exposure to solar ultraviolet radiation.

heat shock proteins
A group of proteins present in all life forms. They are induced when a cell undergoes various types of environmental stresses like heat, cold and oxygen deprivation.

herbicide
Any chemical substance used to destroy plants, esp. weeds, or to check their growth.

hydrocarbon
Any compound containing only hydrogen and carbon.

hydrogen
A flammable, colorless, odorless, gaseous chemical element, the lightest of all known substances.

hydrogen cyanide
A colorless, volatile and extremely poisonous chemical compound.

immunogenic
Causing or producing immunity.

insecticide
Any substance used to kill insects.

ionizing radiation
Electromagnetic or separate particle radiation able to cause ionization.

lumbar-puncture
Insertion of a hollow needle into the lower region of the spinal cord to withdraw cerebrospinal fluid, introduce drugs, etc.

lymphoma
Any form of cancer of the lymph nodes, which help to protect against infection by killing bacteria and neutralizing toxins.

mesothelioma
A tumor of the lining under the lungs; thought to be caused most commonly by the inhalation of asbestos particles.

methylphenidate
A mild central nervous system stimulant prescribed for individuals who have an abnormally high level of activity.

methyl isocyanate
A highly toxic, flammable, colorless liquid, used in the manufacture of pesticides.

microbe
A microscopic organism.

micro-organism
Any microscopic or ultramicroscopic organism.

microwave (radiation)
Short waves of electromagnetic energy that travel at light-speed.

MRSA (Methicillin Resistant Staphylococcus aureus)
A bacteria commonly found on the skin of healthy people.

nuclear fission
The splitting of the nuclei of atoms into two fragments of approximately equal mass, accompanied by conversion of part of the mass into energy; the principle of the atomic bomb.

organophosphate
Any organic compound containing phosphorous, specifically one used as an insecticide, like malathion.

PCBs
Polychlorinated biphenyl, a group of chlorinated isomers of biphenyl, used in the form of a toxic, colorless, odorless, viscous liquid for manufacture of countless products.

penicillin
A group of isomeric, antibiotic compounds obtained from the filtrates of certain molds or produced synthetically.

peripheral/neuritis
Inflammation of nerves, often associated with a degenerative process, accompanied by changes in sensory and motor activity.

polymer acrylamide
Used for water treatment, paper manufacture and in applications requiring water soluble polymers.

plutonium
A radioactive, metallic chemical element found in native uranium ores; its most important isotope, plutonium-239, is used in nuclear weapons and as a reactor fuel.

psychotropic
Having an altering effect on the mind, caused by tranquilizers, hallucinogens, etc.

radioactive
Giving off, or capable of giving off, radiant energy in the form of particles or rays by the spontaneous disintegration of atomic nuclei.

radiology
The science that deals with X-rays and other forms of radiant energy.

sarin
A highly toxic nerve gas, which attacks the central nervous system, bringing on convulsions and death.

silicone/silica
Silicone is a nonmetallic chemical element, more abundant in nature than any other element except oxygen, with which it combines to form silica.

smallpox
An acute, highly contagious virus characterized by fever, vomiting and pustular eruptions that leave pitted scars when healed.

subsonic
Traveling at a velocity below that of sound.

syphilis
An infectious venereal disease. If untreated, it can lead to the degeneration of bones, heart, etc.

thalidomide
A crystalline solid formerly used as a sedative; found to be responsible for birth deformities when taken during pregnancy.

thymus (gland)
A gland in the neck of all vertebrates, involved in the production of lymphocytes, a type of white blood cells.

toxicology
The science of poisons, their effects, antidotes, etc.

trichlorophenoxyacetic acid
A widely-used herbicide.

trichloropyridinol
A major product of chlorpyrifos.

uranium
A very hard, heavy, silvery, moderately malleable, radioactive metallic chemical element.

uranium hexafluoride
A chemical compound consisting of one atom of uranium with six atoms of fluorine.

vivisection
Medical experiments performed on living animals to study the structure and function of living organs and parts, and to investigate the effects of diseases and therapy.

Index

Bibliography

Allen M. Hornblum. *Acres of Skin*. Routledge, New York. 1998.

James H. Jones. *Bad Blood—The Tuskegee Syphilis Experiment*. The Free Press, New York. 1981.

Dominique Lapierre & Javier Moro. *Five Past Midnight in Bhopal*. Scribner, London. 2002.

Dr. George L. Waldbott et al. *Fluoridation—The Great Dilemma*. Coronado Press, Lawrence, Kansas. 1978.

Robin Baker. *Fragile Science—The Reality Behind the Headlines*. Macmillan, London. 2001.

Bjorn Lomborg. *The Skeptical Environmentalist*. Cambridge University Press, UK. 2001.

Geoffrey Cannon. *Superbug—Nature's Revenge*. Virgin Publishing, London. 1995.

Peter Williams & David Wallace. *Unit 731—The Japanese Army's Secret of Secrets*. Hodder & Stoughton, Kent. 1989.

Picture Credits